CAMBRIDGE LIBR

Books of enduring scholarly value

Life Sciences

Until the nineteenth century, the various subjects now known as the life sciences were regarded either as arcane studies which had little impact on ordinary daily life, or as a genteel hobby for the leisured classes. The increasing academic rigour and systematisation brought to the study of botany, zoology and other disciplines, and their adoption in university curricula, are reflected in the books reissued in this series.

History of British Birds

Thomas Bewick (1753–1828) is synonymous with finely crafted wood engravings of the natural world, and his instantly recognisable style influenced book illustration well into the nineteenth century. During his childhood in the Tyne valley, his two obsessions were art and nature. At fourteen, he was apprenticed to the engraver and businessman Ralph Beilby (1743–1817) with whom he later published *A General History of Quadrupeds* (also reissued in this series). The present work, with its text compiled from various sources, was the first practical field guide for the amateur ornithologist, inspiring also artists and writers. Each of the two volumes contains hundreds of illustrations of breathtaking beauty and precision: one for each species, neatly capturing its character in exquisite detail, interspersed with charming vignettes of country life. Volume 1, first published in 1797, covers land birds, including eagles, owls, sparrows and finches.

Cambridge University Press has long been a pioneer in the reissuing of out-of-print titles from its own backlist, producing digital reprints of books that are still sought after by scholars and students but could not be reprinted economically using traditional technology. The Cambridge Library Collection extends this activity to a wider range of books which are still of importance to researchers and professionals, either for the source material they contain, or as landmarks in the history of their academic discipline.

Drawing from the world-renowned collections in the Cambridge University Library and other partner libraries, and guided by the advice of experts in each subject area, Cambridge University Press is using state-of-the-art scanning machines in its own Printing House to capture the content of each book selected for inclusion. The files are processed to give a consistently clear, crisp image, and the books finished to the high quality standard for which the Press is recognised around the world. The latest print-on-demand technology ensures that the books will remain available indefinitely, and that orders for single or multiple copies can quickly be supplied.

The Cambridge Library Collection brings back to life books of enduring scholarly value (including out-of-copyright works originally issued by other publishers) across a wide range of disciplines in the humanities and social sciences and in science and technology.

History of British Birds

VOLUME 1:
CONTAINING THE HISTORY
AND DESCRIPTION OF LAND BIRDS

THOMAS BEWICK

CAMBRIDGE
UNIVERSITY PRESS

University Printing House, Cambridge, CB2 8BS, United Kingdom

Published in the United States of America by Cambridge University Press, New York

Cambridge University Press is part of the University of Cambridge.
It furthers the University's mission by disseminating knowledge in the pursuit of education, learning and research at the highest international levels of excellence.

www.cambridge.org
Information on this title: www.cambridge.org/9781108065405

This edition first published 1797
This digitally printed version 2013

ISBN 978-1-108-06540-5 Paperback

HISTORY

OF

BRITISH BIRDS.

THE FIGURES ENGRAVED ON WOOD BY T. BEWICK.

VOL. I.

CONTAINING THE

HISTORY AND DESCRIPTION OF LAND BIRDS.

NEWCASTLE:

PRINTED BY SOL. HODGSON, FOR BEILBY & BEWICK: SOLD BY THEM, AND G. G. & J. ROBINSON, LONDON.

[*Price* 10*s*. 6*d*. *in Boards*.]

1797.

PREFACE.

TO thoſe who attentively conſider the ſubject of Natural Hiſtory, as diſplayed in the animal creation, it will appear, that though much has been done to explore the intricate paths of Nature, and follow her through all her various windings, much yet remains to be done before the great œconomy is completely developed. Notwithſtanding the laborious and not unſucceſsful inquiries of ingenious men in all ages, the ſubject is far from being exhauſted. Syſtems have been formed and exploded, and new ones have appeared in their ſtead; but, like ſkeletons injudiciouſly put together, they give but an imperfect idea of that order and ſymmetry to which they are intended to be ſubſervient: They have their uſe, but it is the ſkilful practitioner who is chiefly enabled to profit by them; to the leſs informed they appear obſcure and perplexing, and too frequently deter him from the great object of his purſuit.

To inveſtigate, with any tolerable degree of ſucceſs, the more retired and diſtant parts of the animal œconomy, is a taſk of no ſmall difficulty. An inquiry ſo deſireable and ſo eminently uſeful would require the united efforts of many to give it the deſired ſucceſs. Men of leiſure, of all deſcriptions, reſiding in the country, could ſcarcely find a more delightful employment than in attempting to elucidate, from their own obſervations, the various branches of Natural Hiſtory, and in communicating them to others. Something like a ſociety in each county, for the purpoſe of collecting a variety of theſe obſervations, as well as for general correſpondence, would be extremely uſeful and neceſſary: Much might be expected from a combination of this kind extending through every part of the kingdom; a general mode of communication might be thereby eſtabliſhed, in order to aſcertain the changes which are continually taking place, particularly among the feathered tribes; the times of their appearing and diſappearing would be carefully noted; the differences of age, ſex, food, &c. would claim a particular degree of attention, and would be the means of correcting a number of errors which have crept into the works of ſome of the moſt eminent ornithologiſts, and of avoiding the confuſion ariſing from an over-anxious deſire of encreaſing the numbers of each particular kind: But it is reſerved, perhaps, for times of greater tranquillity, when the human mind, undiſturbed by public calamities, ſhall find leiſure to employ itſelf, without interruption, in the purſuit of thoſe objects which enlarge its powers and give dignity to its exertions, to carry into the fulleſt effect a plan for inveſtigations of this ſort.

In this reſpect no author has been more ſucceſsful than the celebrated Count de Buffon:—Deſpiſing the reſtraints which methodical arrangements generally impoſe, he ranges at large through the various walks of Nature, and deſcribes her with a brilliancy of colouring which the moſt lively imagination only could ſuggeſt It muſt, however, be allowed, that in many inſtances, that ingenious philoſopher has overſtepped the bounds of Nature, and, in giving the reins to his own luxuriant fancy, has been too frequently hurried into the wild paths of conjecture and romance. The late Mr White, of Selborne, has added much to the general ſtock of knowledge on this delightful ſubject, by attentively and faithfully recording whatever fell under his own obſervations, and by liberal communications to others.

As far as we could, consistently with the plan laid down in the following work, we have consulted, and we trust with some advantage, the works of these and other Naturalists. In the arrangement of the various classes, as well as in the descriptive part, we have taken as our guide our ingenious countryman, Mr Pennant, to whose elegant and useful labours the world is indebted for a fund of the most rational entertainment, and who will be remembered by every lover of Nature as long as her works have power to charm. The communications with which we have been favoured by those gentlemen who were so good as to notice our growing work, have been generally acknowledged in their proper place; it remains only that we be permitted to insert this testimony of our grateful sense of them.

In a few instances we have ventured to depart from the usual method of classification; by placing the hard-billed birds, or those which live chiefly on seeds, next to those of the Pie kind, there seems to be a more regular gradation downwards, a few anomalous birds, such as the Cuckoo, Hoopoe, Nuthatch, &c. only intervening: The soft-billed birds, or those which subsist chiefly on worms, insects, and such like, are by this means placed all together, beginning with those of the Lark kind. To this we must observe, that, by dividing the various families of birds into two grand divisions, viz. Land and Water, a number of tribes have thereby been included among the latter, which can no otherwise be denominated Water Birds than as they occasionally seek their food in moist places, by small streamlets, or on the sea-shore; such as the Curlew, Woodcock, Snipe, Sandpiper, and many others. These, with such as do not commit themselves wholly to the waters, are thrown into a separate division, under the denomination of Waders. To these we have ventured to remove the Kingfisher, and the Water Ouzel; the former lives entirely on fish, is constantly found on the margins of still waters, and may with greater propriety be denominated a Water Bird than many which come under that description; the latter seems to have no connection with those birds among which it is ufually claffed, its bufinefs being wholly among rapid running ftreams, in which it chiefly delights, and from whence it derives its fupport.

This work, of which the firft volume is all that is now offered to the public, will contain an account of all the various tribes of birds either conftantly refiding in, or occafionally vifiting, our ifland, accompanied with reprefentations of almoft every fpecies, faithfully drawn from Nature, and engraven on wood. It may be proper to obferve, that while one of the Editors of this work was engaged in preparing the engravings, the compilation of the defcriptions was undertaken by the other, fubject, however, to the corrections of his friend, whofe habits had led him to a more intimate acquaintance with this branch of Natural Hiftory: The Compiler, therefore, is anfwerable for the defects which may be found in this part of the undertaking, concerning which he has little to fay, but that it was the production of thofe hours which could be fpared from a laborious employment, and on that account he hopes the feverity of criticifm will be fpared, and that it will be received with that indulgence which has been already experienced on a former occafion.

NEWCASTLE UPON TYNE, *September*, 1797.

INTRODUCTION.

IN no part of the animal creation are the wiſdom, the goodneſs, and the bounty of Providence diſplayed in a more lively manner than in the ſtructure, formation, and various endowments of the feathered tribe. The ſymmetry and elegance diſcoverable in their outward appearance, although highly pleaſing to the ſight, are yet of much greater importance when conſidered with reſpect to their peculiar habits and modes of living, to which they are eminently ſubſervient.

Inſtead of the large head and formidable jaws, the deep capacious cheſt, the brawny ſhoulders, and ſinewy legs of the quadrupeds, we obſerve the pointed beak, the long and pliant neck, the gently ſwelling ſhoulder, the expanſive wings, the tapering tail, the light and bony feet; all which are wiſely calculated to aſſiſt and accelerate their motion through the yielding air. Every part of their frame is formed for lightneſs and buoyancy; their bodies are covered with a ſoft and delicate plumage, ſo diſpoſed as to protect them from the intenſe cold of the atmoſphere through which they paſs; their wings are made of the lighteſt materials, and yet the force with which they ſtrike the air with them is ſo great as to impel their bodies forward with aſtoniſhing rapidity, whilſt the tail ſerves the purpoſe of a rudder to direct them to the different objects of their purſuit. The internal ſtructure of birds is no leſs nicely adapted to the ſame wiſe and uſeful purpoſes; all their bones are light and thin, and all the muſcles, except thoſe which are appropriated to the purpoſe of moving the wings, are extremely delicate and light; the lungs are placed cloſe to the back-bone and ribs, the air entering into them by a communication from the windpipe, paſſes through and is conveyed into a number of membranous cells which lie upon the ſides of the pericardium, and communicate with thoſe of the ſternum. In ſome birds theſe cells are continued down the wings, and extend even to the pinions, thigh bones, and other parts of the body, which can be filled and diſtended with air at the pleaſure of the animal.

The extreme ſingularity of this almoſt univerſal diffuſion of air through the bodies of birds naturally excited a ſtrong deſire to diſcover the intention of Nature in producing a conformation ſo extraordinary. The ingenious Mr Hunter imagined that it might be intended to aſſiſt the animal in the act of flying, by increaſing its bulk and ſtrength without adding to its weight. This opinion was corroborated by conſidering, that the feathers of birds, and particularly thoſe of the wings, contain a great quantity of air. In oppoſition to this he informs us, that the Oſtrich, which does not fly, is nevertheleſs provided with air cells difperfed through its body; that the Woodcock, and fome other flying birds, are not fo liberally fupplied with thefe cells; yet, he elfewhere obferves, that it may be laid down as a general rule, that in birds of the higheft and longeft flights, as Eagles, this extenfion or diffufion of air is carried further than in others; and, with regard to the Oftrich, though it is deprived of the power of flying, it runs with amazing rapidity, and confequently requires fimilar refources of air. It feems therefore to be proved, evidently, that this general diffufion of air through the bodies of birds is of infinite ufe to them, not only in their long and laborious flights, but likewife in preventing their refpiration from being ftopped or interrupted by the rapidity of their motion through a refifting medium. Were it poffible for man to move with the fwiftnefs of a Swallow, the actual refiftance of the air, as he is not provided with internal refervoirs fimilar to thofe of birds, would foon fuffocate him.*

Birds may be diftinguifhed, like quadrupeds, into two kinds or claffes, granivorous and carnivorous; like quadrupeds too, there are fome that hold a middle nature, and partake of both. Granivorous birds are furnifhed with larger inteftines, and proportionally longer than thofe of the carnivorous kinds. Their food, which confifts of grain of various kinds, is conveyed whole and entire into the firft ftomach or craw, where it undergoes a partial dilution by a liquor fecreted from glands fpread over its furface; it is then received into another fpecies of ftomach, where it is further diluted; after which it is tranfmitted into the gizzard, or true ftomach, confifting of two very ftrong mufcles covered externally with a tendinous fubftance, and lined with a thick membrane of prodigious power and ftrength; in this place the food is completely triturated, and rendered fit for the operation of the gaftric juices. The extraordinary powers

* May not this univerfal diffufion of air through the bodies of birds account for the fuperior heat of this clafs of animals? The feparation of oxygen from refpirable air, and its mixture with the blood, by means of the lungs, being fuppofed by the ingenious Dr Crawford to be the efficient caufe of animal heat.

of the gizzard in comminuting the food, ſo as to prepare it for digeſtion, would exceed all credibility, were they not ſupported by incontrovertible facts founded upon experiments. In order to aſcertain the ſtrength of theſe ſtomachs, the ingenious Spalanzani made the following curious and very intereſting experiments:—Tin tubes, full of grain, were forced into the ſtomachs of Turkies, and after remaining twenty hours, were found to be broken, compreſſed, and diſtorted in the moſt irregular manner.* In proceeding further, the ſame author relates, that the ſtomach of a Cock, in the ſpace of twenty-four hours, broke off the angles of a piece of rough jagged glaſs, and upon examining the gizzard, no wound or laceration appeared. Twelve ſtrong needles were firmly fixed in a ball of lead, the points of which projected about a quarter of an inch from the ſurface; thus armed, it was covered with a caſe of paper, and forced down the throat of a Turkey; the bird retained it a day and a half, without ſhewing the leaſt ſymptom of uneaſineſs; the points of all the needles were broken off cloſe to the ſurface of the ball, except two or three, of which the ſtumps projected a little. The ſame author relates another experiment, ſeemingly ſtill more cruel: He fixed twelve ſmall lancets, very ſharp, in a ſimilar ball of lead, which was given in the ſame manner to a Turkey-cock, and left eight hours in the ſtomach; at the expiration of which the organ was opened, but nothing appeared except the naked ball, the twelve lancets having been broken to pieces, the ſtomach remaining perfectly ſound and entire. From theſe curious and well-atteſted facts we may conclude, that the ſtones ſo often found in the ſtomachs of many of the feathered tribe are highly uſeful in comminuting the grain and other hard ſubſtances which conſtitute their food. "The ſtones," ſays the celebrated Dr Hunter, "aſſiſt in grinding down the grain, and, by ſeparating its parts, allow the gaſtric juices to come more readily into contact with it." Thus far the concluſion coincides with the experiments which have been juſt related. We may

* Spalanzani's Diſſertation, vol. 1, page 12.

obſerve ſtill farther, that the ſtones thus taken into the ſtomachs of birds are ſeldom known to paſs with the fæces, but being ground down and ſeparated by the powerful action of the gizzard, are mixed with the food, and, no doubt, contribute very much to the health as well as nutriment of the animal.

Granivorous birds partake much of the nature and diſpoſition of herbivorous quadrupeds. In both, the number of their ſtomachs, the length and capacity of their inteſtines, and the quality of their food, are very ſimilar; they are likewiſe both diſtinguiſhed by the gentleneſs of their tempers and manners: Contented with the ſeeds of plants, with fruits, inſects, and worms, their chief attention is directed to procuring food, hatching and rearing their offspring, and avoiding the ſnares of men, and the attacks of birds of prey and other rapacious animals. They are a mild and gentle race, and are in general ſo tractable as eaſily to be domeſticated. Man, ever attentive and watchful to every thing conducive to his intereſt, has not failed to avail himſelf of theſe diſpoſitions, and has judiciouſly ſelected, from the numbers which every way ſurround him, thoſe which are moſt prolific, and conſequently the moſt profitable: Of theſe the Hen, the Gooſe, the Turkey, and the Duck are the moſt conſiderable, and form an inexhauſtible ſtore of rich, wholeſome, and nutritious food.

Carnivorous birds are diſtinguiſhed by thoſe endowments and powers with which they are furniſhed by Nature for the purpoſe of procuring their food: They are provided with wings of great length, the muſcles which move them being proportionally large and ſtrong, whereby they are enabled to keep long upon the wing in ſearch of their prey; they are armed with ſtrong hooked bills, ſharp and formidable claws; they have alſo large heads, ſhort necks, ſtrong and brawny thighs, and a ſight ſo acute and piercing, as to enable them to view their prey from immeaſureable heights in the air, upon which they dart with inconceiveable ſwiftneſs and undeviating aim; their ſtomachs are ſmaller than thoſe of the granivorous kinds, and their inteſtines are much ſhorter. The analogy between the ſtructure of rapacious birds and carnivorous quadrupeds is obvious; both of them are provided with weapons which indicate deſtruction and rapine, their manners are fierce and unſocial, and they ſeldom herd together in flocks like the inoffenſive granivorous tribes. When not on the wing, rapacious birds retire to the tops of ſequeſtered rocks, or the depths of extenſive foreſts, where they conceal themſelves in ſullen and gloomy ſolitude. Thoſe which feed on carrion are endowed with a ſenſe of ſmelling ſo exquiſite, as to enable them to ſcent dead and putrid carcaſes at aſtoniſhing diſtances.

Beſide theſe great diviſions of birds into granivorous and rapacious kinds, there are numerous other tribes, to whom Nature has given ſuitable organs, adapted to their peculiar habits and modes of living. Like amphibious animals, a great variety of birds live chiefly in the water, and feed on fiſhes, inſects, and other aquatic productions: To enable them to ſwim and dive in queſt of food, their toes are connected by broad membranes or webs, with which they ſtrike the water, and are driven forward with great force. The ſeas, the lakes, and rivers abound with innumerable ſwarms of birds of various kinds, all which find an abundant ſupply in the immeaſurable ſtores with which the watery world is every where ſtocked. There are other tribes of aquatic birds, frequenting marſhy places and the margins of lakes and rivers, which ſeem to partake of a middle nature between thoſe which live wholly on land, and thoſe which are entirely occupied in waters: Some of theſe feed on fiſhes and reptiles; others derive nouriſhment by thruſting their long bills into ſoft and muddy ſubſtances, where they find worms, the eggs of inſects, and other nutritious matter; they do not ſwim, but wade in queſt of food, for which purpoſe Nature has provided them with long legs, bare of feathers even above the knees; their toes are not connected by webs, like thoſe of the ſwimmers, but are only partially furniſhed with membranaceous appendages, which are juſt ſufficient to ſupport them on the ſoft and doubtful ground which they are accuſtomed to frequent:— Moſt of theſe kinds have very long necks and bills, to enable them to ſearch for and find their concealed food. To theſe tribes belong the Crane, the Heron, the Bittern, the Stork, the Spoonbill, the Woodcock, the Snipe, and many others.

Without the means of conveying themſelves with great ſwiftneſs from one place to another, birds could not eaſily ſubſiſt: The food which Nature has ſo bountifully provided for them is ſo irregularly diſtributed, that they are obliged to take long journies to diſtant parts in order to gain the neceſſary ſupplies; at one time it is given in great abundance; at another it is adminiſtered with a very ſparing hand; and this is one cauſe of thoſe migrations ſo peculiar to the feathered tribe. Beſides the want of food, there are two other cauſes of migration, viz. the want of a proper temperature of air, and a convenient ſituation for the great work of breeding and rearing their young. Such birds as migrate to great diſtances are alone denominated *birds of paſſage;* but moſt birds are, in ſome meaſure, birds of paſſage, although they do not migrate to places remote from their former habitations. At particular times of the year moſt birds remove from one country to another, or from the more inland diſtricts toward the ſhores: The times of theſe migrations or flittings are obſerved with the moſt aſtoniſhing order and punctuality; but the ſecrecy of their departure and the ſuddenneſs of their re-appearance have involved the ſubject of migration in general in great difficulties. Much of this difficulty ariſes from our not being able to account for the means of ſubſiſtence during the long flights of many of thoſe birds, which are obliged to croſs immenſe tracts of water before they arrive at the places of their deſtination: Accuſtomed to meaſure diſtances by the ſpeed of thoſe animals with which we are well acquainted, we are apt to overlook the ſuperior velocity with which birds are carried forward in the air, and the eaſe with which they continue their exertions for a much longer time than can be done by the ſtrongeſt quadruped.

Our ſwifteſt horſes are ſuppoſed to go at the rate of a mile in ſomewhat leſs than two minutes, and we have one inſtance on record of a horſe being tried, which went at the rate of nearly a mile in one minute, but that was only for the ſmall ſpace of a ſecond of time.* In this and ſimilar inſtances we find, that an uncommon degree of exertion was attended with its uſual conſequences, debility, and a total want of power to continue it to the ſame extent; but the caſe is very different with birds, their motions are not impeded by the ſame cauſes, they glide through the air with a quickneſs ſuperior to that of the ſwifteſt quadruped, and they can continue on the wing with the ſame ſpeed for a conſiderable length of time. Now, if we can ſuppoſe a bird to go at the rate of only half a mile in a minute, for the ſpace of twenty-four hours, it will have gone over, in that time, an extent of more than ſeven hundred miles, which is ſufficient to account for almoſt the longeſt migration; but if aided by a favourable current of air, there is reaſon to ſuppoſe that the ſame journey may be performed in a much ſhorter ſpace of time. To theſe obſervations we may add, that the ſight of birds is peculiarly quick and piercing; and from the advantage they poſſeſs in being raiſed to conſiderable heights in the air, which is well known to be the caſe with the Stork, Bittern, and other kinds of birds, they are enabled, with a ſagacity peculiar to inſtinctive knowledge, to diſcover the route they are to take, from the appearance of the atmoſphere, the clouds, the direction of the winds, and other cauſes; ſo that, without having recourſe to improbable modes, it is eaſy to conceive, from the velocity of their ſpeed alone, that moſt birds may tranſport themſelves to countries laying at great diſtances, and acroſs vaſt tracts of ocean.

The following obſervations from Cateſby are very applicable, and will conclude our remarks on this head: "The manner of "their journeyings may vary according as the ſtructure of their "bodies enables them to ſupport themſelves in the air. Birds "with ſhort wings, ſuch as the Redſtart, Black-cap, &c. may "paſs by gradual and ſlower movements; and there ſeems no "neceſſity for a precipitate paſſage, as every day affords an in-

* See Hiſtory of Quadrupeds, page 6, 3d edition.

" crease of warmth, and a continuance of food. It is probable " these itinerants may perform their journey in the night-time, " in order to avoid ravenous birds, and other dangers which " day-light may expose them to. The flight of the smaller " birds of passage across the seas has, by many, been considered " as wonderful, and especially with regard to those with short " wings, among which Quails seem by their structure little a- " dapted for long flights; nor are they ever seen to continue " on the wing for any length of time, and yet their ability for " such flights cannot be doubted. The coming of these birds " is certain and regular from every year's experience, but the " cause and manner of their departure have not always been so " happily accounted for; in short, all we know of the matter " ends in this observation;—that Providence has created a " great variety of birds and other animals with constitutions " and inclinations adapted to their several wants and necessities, " as well as to the different degrees of heat and cold in the se- " veral climates of the world, whereby no country is destitute " of inhabitants, and has given them appetites for the produc- " tions of those countries whose temperature is suited to their " nature, as well as knowledge and ability to seek and find " them out."

The migration of the Swallow tribe has been noticed by almost every writer on the natural history of birds, and various opinions have been formed respecting their disappearance, and the state in which they subsist during that interval. Some Naturalists suppose that they do not leave this island at the end of autumn, but that they lie in a torpid state, till the beginning of summer, in the banks of rivers, in the hollows of decayed trees, in holes and crevices of old buildings, in sand banks, and the like: Some have even asserted that Swallows pass the winter immersed in the waters of lakes and rivers, where they have been found in clusters, mouth to mouth, wing to wing, foot to foot, and that they retire to these places in autumn, and creep down the reeds to their subaqueous retreats. In support of this opinion, Mr Klein very gravely asserts, on the credit of some countrymen, that Swallows ſometimes aſſembled in numbers, clinging to a reed till it broke, and ſunk with them to the bottom; that their immerſion was preceded by a ſong or dirge, which laſted more than a quarter of an hour; ſometimes they laid hold of a ſtraw with their bills, and plunged down in ſociety; and that others formed a large maſs, by clinging together by the feet, and in this manner committing themſelves to the deep. It requires no great depth of reaſoning to refute ſuch palpable abſurdities, or to ſhew the phyſical impoſſibility of a body, ſpecifically lighter than water, employing another body lighter than itſelf for the purpoſe of immerſion: But, admitting the poſſibility of this curious mode of immerſion, it is by no means probable that Swallows, or any other animal, in a torpid ſtate, can exiſt for any length of time in an element to which they have never been accuſtomed, and are beſides totally unprovided by Nature with organs ſuited to ſuch a mode of ſubſiſtence.

The celebrated Mr John Hunter informs us, "that he had diſſected many Swallows, but found nothing in them different from other birds as to the organs of reſpiration;" and therefore concludes that it is highly abſurd to ſuppoſe, that terreſtrial animals can remain any long time under water without drowning. It muſt not however be denied, that Swallows have been ſometimes found in a torpid ſtate during the winter months; but ſuch inſtances are by no means common, and will not ſupport the inference, that, if any of them can ſurvive the winter in that ſtate, the whole ſpecies is preſerved in the ſame manner.* That other

* There are various inſtances on record, which bear the ſtrongeſt marks of veracity, of Swallows having been taken out of water, and of their having been ſo far recovered by warmth as to exhibit evident ſigns of life, ſo as even to fly about for a ſhort ſpace of time. But whilſt we admit the fact, we are not inclined to allow the concluſion generally drawn from it, viz. that Swallows, at the time of their diſappearance, frequently immerſe themſelves in ſeas, lakes, and rivers, and at the proper ſeaſon emerge and re-aſſume the ordinary functions of life and animation; for, it ſhould be obſerved, that in thoſe inſtances, which have been the beſt authenticated, [See Forſter's Tranſlation of Kalm's Travels into North America, p. 140—note.] it appears, that the Swallows ſo taken up

birds have been found in a torpid ſtate may be inferred from the following curious fact, which was communicated to us by a gentleman who ſaw the bird, and had the account from the perſon who found it. A few years ago, a young Cuckoo was found in the thickeſt part of a cloſe whin buſh; when taken up it preſently diſcovered ſigns of life, but was quite deſtitute of feathers; being kept warm, and carefully fed, it grew and recovered its coat of feathers: In the ſpring following it made its eſcape, and in flying acroſs the river Tyne it gave its uſual call. We have obſerved a ſingle Swallow ſo late as the latter end of October. Mr White, in his Natural Hiſtory of Selborne, mentions having ſeen a Houſe Martin flying about in November, long after the general migration had taken place. Many more inſtances might be given of ſuch late appearances, which, added to the well-authenticated accounts of Swallows having been actually found in a torpid ſtate, leave us no room to doubt, that ſuch young birds as have been late hatched, and conſequently not ſtrong enough to undertake a long voyage to the coaſt of Africa, are left behind, and remain concealed in hiding places till the return of ſpring: On the other hand, that actual migrations of the Swallow tribes do take place, has been fully proved from a variety of well-atteſted

were generally found entangled amongſt reeds and ruſhes, by the ſides, or in the ſhalloweſt parts of the lakes or rivers where they happened to be diſcovered, and that having been brought to life ſo far as to fly about, they all of them died in a few hours after. From the facts thus ſtated we would infer, that at the time of the diſappearance of Swallows, the reedy grounds by the ſides of rivers and ſtanding waters are generally dry, and that theſe birds, eſpecially the later hatchings, which frequent ſuch places for the ſake of food, retire to them at the proper ſeaſon, and lodge themſelves among the roots, or in the thickeſt parts of the rank graſs which grows there; that during their ſtate of torpidity they are liable to be covered with water, from the rains which follow, and are ſometimes waſhed into the deeper parts of the lake or river where they have been accidentally taken up; and that probably the tranſient ſigns of life which they have diſcovered on ſuch occaſions, have given riſe to a variety of vague and improbable accounts of their immerſion, &c.

facts, moſt of which have been taken from the obſervations of navigators who have been eye-witneſſes of their flights, and whoſe ſhips have ſometimes afforded a reſting-place to the weary travellers.

To the many on record we ſhall add the following, which we received from a very ſenſible maſter of a veſſel, who, whilſt he was ſailing early in the ſpring between the iſlands of Minorca and Majorca, ſaw great numbers of Swallows flying northward, many of whom alighted on the rigging of the ſhip in the evening, but diſappeared before morning. After all our inquiries into this branch of natural œconomy, much yet remains to be known, and we may conclude, in the words of the ingenious Mr White, "that whilſt we obſerve with delight with " how much ardour and punctuality thoſe little birds obey " the ſtrong impulſe towards migration or hiding, imprinted " on their minds by their great Creator, it is with no ſmall de- " gree of mortification that we reflect, that after all our pains " and inquiries, we are not yet quite certain to what regions " they do migrate, and are ſtill farther embarraſſed to find that " ſome do not actually migrate at all.

" Amuſive birds! ſay where your hid retreat,
" When the froſt rages, and the tempeſts beat;
" Whence your return, by ſuch nice inſtinct led,
" When Spring, ſweet ſeaſon, lifts her bloomy head?
" Such baffled ſearches mock man's prying pride,
" The GOD of NATURE is your ſecret guide!"

Moſt birds, at certain ſeaſons, live together in pairs; the union is formed in the ſpring, and generally continues whilſt the united efforts of both are neceſſary in forming their temporary habitations, and in rearing and maintaining their offſpring. Eagles and other birds of prey continue their attachment for a much longer time, and ſometimes for life. The neſts of birds are conſtructed with ſuch exquiſite art, as to exceed the utmoſt exertion of human ingenuity to imitate them. Their mode of building, the materials they make uſe of, as well as the ſituations they ſelect, are as various as the different kinds of birds, and are all admirably adapted to their ſeveral wants and neceſſities. Birds of the ſame ſpecies, whatever region of the globe they inhabit, collect the ſame materials, arrange them in the ſame manner, and make choice of ſimilar ſituations for fixing the places of their temporary abodes. To deſcribe minutely the different kinds of neſts, the various ſubſtances of which they are compoſed, and the judicious choice of ſituations, would ſwell this part of our work much beyond its due bounds. Every part of the world furniſhes materials for the aerial architects; leaves and ſmall twigs, roots and dried graſs, mixed with clay, ſerve for the external; whilſt moſs, wool, fine hair, and the ſofteſt animal and vegetable downs, form the warm internal part of theſe commodious dwellings. The following beautiful lines from Thomſon are highly deſcriptive of the buſy ſcene which takes place during the time of nidification.

> " ———— Some to the holly hedge
> " Neſtling repair, and to the thicket ſome;" &c. &c.*

After the buſineſs of incubation is over, and the young are ſufficiently able to provide for themſelves, the neſts are always abandoned by the parents, excepting thoſe of the Eagle kind.

The various gifts and endowments which the great Author of Nature has ſo liberally beſtowed upon his creatures in general, demand, in a peculiar manner, the attention of the curious Naturaliſt; amongſt the feathered tribe in particular there is much room, in this reſpect, for minute and attentive inveſtigation. In purſuing our inquiries into that ſyſtem of œconomy, by which every part of Nature is upheld and preſerved, we are ſtruck with wonder in obſerving the havock and deſtruction which every where prevail throughout the various orders of beings inhabiting the earth. Our humanity is intereſted in that law of Nature, which devotes to deſtruction myriads of creatures to ſupport and continue the exiſtence of others; but, although

* See Thomſon's Seaſons—Spring.

it is not allowed us to unravel the myſterious workings of Nature through all her parts, or unfold her deep deſigns, we are, nevertheleſs, ſtrongly led to the conſideration of the means by which individuals, as well as ſpecies, are preſerved and multiplied. The weak are frequently enabled to elude the purſuits of the ſtrong, by flight or ſtratagem; ſome are ſcreened from the purſuit of their enemies, by an arrangement of colours happily aſſimilated to the places which they moſt frequent, and where they find either food or repoſe; thus the Wryneck is ſcarcely to be diſtinguiſhed from the bark of the tree on which it feeds, or the Snipe from the ſoft and moſſy ground by the ſprings of water which it frequents; the Great Plover finds its greateſt ſecurity in ſtony places, to which its colours are ſo nicely adapted, that the moſt exact obſerver may be very eaſily deceived.

The attentive Ornithologiſt will not fail to diſcover numerous inſtances of this kind, ſuch as the Partridge, Plover, Quail, &c. Some are indebted to the brilliancy of their colours as the means of alluring their prey; of this the Kingfiſher is a remarkable inſtance, and deſerves to be particularly noticed. This beautiful bird has been obſerved, in ſome ſequeſtered places, near the edge of a rivulet, expoſing the vivid colours of its breaſt to the full rays of the ſun, and fluttering with expanded wings over the ſmooth ſurface of the water; the fiſh, attracted by the brightneſs and ſplendour of the appearance, are detained whilſt the wily bird darts down upon them with unerring certainty. We do not ſay that the mode of taking fiſh by torch-light has been derived from this practiſed by the Kingfiſher, but every one muſt be ſtruck by the ſimilarity of the means. Others, again, derive the ſame advantage from the ſimplicity of their exterior appearance; of this the Heron will ſerve as an example. He may frequently be ſeen ſtanding motionleſs by the edge of a piece of water, waiting patiently the approach of his prey, which he never fails to ſeize as ſoon as it comes within reach of his long neck; he then reaſſumes his former poſition, and continues to wait with the ſame patient attention as before.

Moſt of the ſmaller birds are ſupported, eſpecially when

young, by a profuſion of caterpillars, ſmall worms, and inſects, with which every part of the vegetable world abounds; which is by this means preſerved from total deſtruction, contrary to the commonly received opinion, that birds, particularly Sparrows, do much miſchief in deſtroying the labours of the gardener and the huſbandman. It has been obſerved, " that a ſingle pair of Sparrows, during the time they are feeding their young, will deſtroy about four thouſand caterpillars weekly; they likewiſe feed their young with butterflies and other winged inſects, each of which, if not deſtroyed in this manner, would be productive of ſeveral hundreds of caterpillars." Swallows are almoſt continually upon the wing, and in their curious winding flights deſtroy immenſe quantities of flies and other inſects which are continually floating in the air, and which, if not deſtroyed by theſe birds, would render it unfit for the purpoſes of life and health. That active little bird, the Tomtit, which has generally been ſuppoſed hoſtile to the young and tender buds which appear in the ſpring, when attentively obſerved, may be ſeen running up and down amongſt the branches, and picking the ſmall worms which are concealed in the bloſſoms, and which would effectually deſtroy the fruit. As the ſeaſon advances, various other ſmall birds, ſuch as the Redbreaſt, Wren, Winter Fauvette or Hedge-ſparrow, Whitethroat, Redſtart, &c. are all engaged in the ſame uſeful work, and may be obſerved examining every leaf, and feeding upon the inſects which they find beneath them. —Theſe are a few inſtances of that ſuperintending providential care, which is continually exerted in preſerving the various ranks and orders of beings in the ſcale of animated Nature; and although it is permitted that myriads of individuals ſhould every moment be deſtroyed, not a ſingle ſpecies is loſt, but every link of the great chain remains unbroken

Great Britain produces a more abundant variety of birds than moſt northern countries, owing to the various condition of our lands, from the higheſt ſtate of cultivation to that of the wildeſt, moſt mountainous, and woody. The great quantities of berries and other kinds of fruit produced in our hedges, heaths, and plantations, bring ſmall birds in great numbers, and birds of prey in conſequence: Our ſhores, and the numerous little iſlands adjacent to them, afford ſhelter and protection to an infinite variety of almoſt all kinds of water fowl. To enumerate the various kinds of birds that viſit this iſland annually will not, we preſume, be unacceptable to our readers, nor improper in this part of our work. The following are ſelected chiefly from Mr White's Natural Hiſtory of Selborne, and are arranged nearly in the order of their appearing.

1	Wryneck, - - - - - - -	Middle of March
2	Smalleſt Willow Wren, - -	Latter end of ditto
3	Houſe Swallow, - - - - -	Middle of April
4	Martin, - - - - - - -	Ibid
5	Sand Martin, - - - - -	Ibid
6	Black-cap, - - - - - - -	Ibid
7	Nightingale, - - - - -	Beginning of April
8	Cuckoo, - - - - - - -	Middle of ditto
9	Middle Willow Wren, - -	Ibid
10	White-throat, - - - - - -	Ibid
11	Redſtart, - - - - - - -	Ibid
12	Great Plover or Stone Curlew,	End of March
13	Graſshopper Lark, - - -	Middle of April
14	Swift, - - - - - - - -	Latter end of ditto
15	Leſſer Reed Sparrow, - -	
16	Corncrake or Land Rail, -	
17	Largeſt Willow Wren, - -	End of April
18	Fern Owl, - - - - - - -	Latter end of May
19	Flycatcher, - - - - - -	Middle of ditto.*

Moſt of the ſoft-billed birds feed on inſects, and not on grain or ſeeds, and therefore uſually retire before winter; but the following, though they eat inſects, remain with us during the whole year, viz. The Redbreaſt, Winter Fauvette, and Wren,

* This is the lateſt ſummer bird of paſſage.

which frequent out-houfes and gardens, and eat fpiders, fmall worms, crumbs, &c. The Pied, the Yellow, and the Grey Wagtail, which frequent the heads of fprings, where the waters feldom freeze, and feed on the aureliæ of infects ufually depofited there: Befides thefe, the Whinchat, the Stonechatter, and the Golden-crefted Wren, are feen with us during the winter; the latter, though the leaft of all the Britifh birds, is very hardy, and can endure the utmoft feverity of our winters. The White rump, though not common, fometimes ftays the winter with us.—Of the winter birds of paffage, the following are the principal, viz.

1 The Redwing or Wind Thrufh.

2 The Fieldfare.—[Both thefe arrive in great numbers about Michaelmas, and depart about the end of February, or beginning of March.]

3 The Hooded or Sea Crow vifits us in the beginning of winter, and departs with the Woodcock.

4 The Woodcock appears about Michaelmas, and leaves us about the beginning of March.

5 Snipes are confidered by Mr White as birds of paffage, though he acknowledges that they frequently breed with us. Mr Pennant remarks, that their young are fo frequently found in Britain, that it may be doubted whether they ever entirely leave this ifland.

6 The Judcock or Jack Snipe.

7 The Wood Pigeon: Of the precife time of its arrival we are not quite certain, but fuppofe it may be fome time in April, as we have feen them in the north at that time.

8 The Wild Swan frequents the coafts of this ifland in large flocks, but is not fuppofed to breed with us: It has been chiefly met with in the northern parts, and is faid to arrive at Lingey, one of the Hebrides, in October, and remains there till March, when it retires more northward to breed.

9 The Wild Goofe paffes fouthward in October, and returns northward in April.

With regard to the Duck kind in general, they are moſtly birds of paſſage. Mr Pennant obſerves, " Of the numerous " ſpecies that form this genus, we know of no more than five " that breed here, viz the Tame Swan, the Tame Gooſe, the " Shield Duck, the Eider Duck, and a very ſmall number of " the Wild Ducks: The reſt contribute to form that amazing " multitude of water fowl that annually repair from moſt parts " of Europe to the woods and lakes of Lapland and other arc-" tic regions, there to perform the functions of incubation and " nutrition in full ſecurity. They and their young quit their re-" treats in September, and diſperſe themſelves over Europe. " With us they make their appearance in the beginning of " October, circulate firſt round our ſhores, and, when compel-" led by ſevere froſt, betake themſelves to our lakes and ri-" vers."—In winter the Bernacles and Brent Geeſe appear in vaſt flocks on the north-weſt coaſt of Britain, and leave us in February, when they migrate as far as Lapland, Greenland, or Spitzbergen.

The Solon Geeſe or Gannets are birds of paſſage; their firſt appearance is in March, and they continue till Auguſt or September. The Long-legged Plover and Sanderling viſit us in winter only; and it is worthy of remark, that every ſpecies of the Curlews, Woodcocks, Sandpipers, and Plovers, which forſake us in the ſpring, retire to Sweden, Poland, Pruſſia, Norway, and Lapland to breed, and return to us as ſoon as the young are able to fly; the froſts, which ſet in early in thoſe countries, depriving them totally of ſubſiſtence.

Beſides theſe, there is a great variety of birds which perform partial migrations, or flittings, from one part of the country to another. During hard winters, when the ſurface of the earth is covered with ſnow, many birds, ſuch as Larks, Snipes, &c. withdraw from the inland parts of the country towards the ſea-ſhores in queſt of food; others, as the Wren, the Redbreaſt, and a variety of ſmall birds, quit the fields, and approach the habitations of men. The Bohemian Chatterer, the Groſbeak, and the Croſsbill, are only occaſional viſitors, and obſerve no regular times in making their appearance: Great numbers of the former were taken in the county of Northumberland the latter end of the years 1789 and 1790, before which they had not been obſerved ſo far ſouth as that county, and ſince that time have never been ſeen there.

The ages of birds are various, and do not ſeem to bear the ſame proportion to the time of acquiring their growth as has been remarked with regard to quadrupeds. Moſt birds acquire their full dimenſions in a few months, and are capable of propagation the firſt ſummer after they are hatched. In proportion to the ſize of their bodies, birds are much more vivacious, and live longer, than either man or quadrupeds: Notwithſtanding the difficulties which ariſe in aſcertaining the ages of birds, there are inſtances of great longevity in many of them. Geeſe and Swans have been known to attain the age of one hundred or upwards; Ravens are very long-lived birds, and are ſaid ſometimes to exceed a century; Eagles are ſuppoſed to arrive at a great age; Pigeons are known to live more than twenty years; and even Linnets and other ſmall birds have been kept in cages from fifteen to twenty years.

To the practical Ornithologiſt there ariſes a conſiderable gratification in being able to diſcern the diſtinguiſhing characters of birds as they appear at a diſtance, whether at reſt, or during their flight; for not only every ſpecies has ſomething peculiar to itſelf, but each genus has its own appropriate marks, upon which a judicious obſerver may diſcriminate with almoſt unerring certainty. Of theſe, the various modes of flight afford the moſt certain and obvious means of diſtinction, and ſhould be noted with the moſt careful attention. From the bold and lofty flight of the Eagle, to the ſhort and ſudden flittings of the Sparrow or the Wren, there is an ample field for the curious inveſtigator of Nature, on which he may dwell with inexpreſſible delight, tracing the various movements of the feathered nations which every where preſent themſelves to his view. The notes, or, as it may with more propriety be called, the language, of birds, whereby they are enabled to expreſs, in no inconſiderable degree, their various paſſions, wants, and feelings, muſt be particularly noticed:* The great power of their voice, by which they can communicate their ſentiments and intentions to each other, and by that means are able to act by mutual concert, added to that of the wing, by which they can remove from place to place with inconceivable celerity and diſpatch, is peculiar to the feathered tribes; it gives them a decided ſuperiority over every ſpecies of quadrupeds, and affords them the greateſt means of ſafety from thoſe attacks to which their weakneſs would otherwiſe expoſe them. The ſocial inſtinct among birds is peculiarly lively and intereſting, and likewiſe proves an effectual means of preſervation from the various arts which are made uſe of to circumvent and deſtroy them. Individuals may periſh, and the ſpecies may ſuffer a diminution of its numbers; but its inſtincts, habits, and œconomy remain entire.

* White's Selborne.

CONTENTS OF THE FIRST VOLUME.

HISTORY

OF

BRITISH BIRDS.

VOL. I.

BRITISH BIRDS.

BIRDS OF PREY.

RAPACIOUS birds, or thoſe which ſubſiſt chiefly on fleſh, are much leſs numerous than ravenous quadrupeds; and it ſeems wiſely provided by nature, that their powers ſhould be equally confined and limited as their numbers; for if, to the rapid flight and penetrating eye of the Eagle, were joined the ſtrength and voracious appetite of the Lion, the Tiger, or the Glutton, no artifice could evade the one, and no ſpeed could eſcape the other.

The characters of birds of the ravenous kind are particularly ſtrong, and eaſily to be diſtinguiſhed; the formidable talons, the large head, the ſtrong and crooked beak, indicate their ability for rapine and carnage; their diſpoſitions are fierce, and their nature untractable; unſociable and cruel, they avoid the haunts of civilization, and retire to the moſt melancholy and wild receſſes of nature, where they can enjoy, in gloomy ſolitude, the effects of their depredatory excurſions. The fierceneſs of their nature extends even to their young, which they drive from the neſt at a very early period; the difficulty of procuring a conſtant ſupply of food for them ſometimes overcomes the feelings of parental affection; and they have been known to deſtroy them in the fury of diſappointed hunger. Different from all other kinds, the female of birds of prey is larger and ſtronger than the male: naturaliſts have puzzled themſelves to aſſign the reaſon of this extraordinary property, but the final cauſe at leaſt is obvious:—The care of rearing her young being ſolely intruſted to the female, nature has furniſhed her with more ample powers to provide for her own wants and thoſe of her offspring.

This formidable tribe conſtitutes the firſt order among the genera of birds. Thoſe of our own country conſiſt only of two kinds, viz. the Falcon and the Owl.—We ſhall begin with the former.

THE FALCON TRIBE.

The numerous families of which this kind is compoſed, are found in almoſt every part of the world, from the frigid to the torrid zone; they are divided into various claſſes or tribes, conſiſting of Eagles, Kites, Buzzards, Hawks, &c. and are readily known by the following diſtinguiſhing characteriſtics:

The bill is ſtrong, ſharp, and much hooked, and is furniſhed with a naked ſkin or cere ſituated at the baſe, in which are placed the noſtrils; the head and neck are well clothed with feathers, which ſufficiently diſtinguiſh it from every one of the vulture kind; the legs and feet are ſcaly, claws large and ſtrong, much hooked, and very ſharp: Birds of this ſpecies are alſo diſtinguiſhed by their undaunted courage, and great activity. Buffon, ſpeaking of the Eagle, compares it with the Lion, and aſcribes to it the magnanimity, the ſtrength, and the forbearance of that noble quadruped. The Eagle deſpiſes ſmall animals, and diſregards their inſults; he ſeldom devours the whole of his prey, but, like the Lion, leaves the fragments to other animals; though famiſhed with hunger, he diſdains to feed on carrion. The eyes of the Eagle have the glare of thoſe of the Lion, and are nearly of the ſame colour; the claws are of the ſame ſhape, and the cry of both is powerful and terrible; deſtined for war and plunder, they are equally fierce, bold, and untractable. Such is the reſemblance which that ingenious and fanciful writer has pictured of theſe two noble animals; the characters of both are ſtriking and prominent, and hence the Eagle is ſaid to extend his dominion over the birds, as the Lion over the quadrupeds.

The ſame writer alſo obſerves, that in a ſtate of nature, the Eagle never engages in a ſolitary chace but when the female is confined to her eggs or her young: at this ſeaſon the return of the ſmaller birds affords plenty of prey, and he can with eaſe provide for the ſuſtenance of himſelf and his mate: at other times they unite their exertions, and are always ſeen cloſe together, or at a ſhort diſtance from each other. Thoſe who have an opportunity of obſerving their motions, ſay, that the one beats the buſhes, whilſt the other, perched on an eminence, watches the eſcape of the prey. They often ſoar out of the reach of human ſight; and notwithſtanding the immenſe diſtance, their cry is ſtill heard, and then reſembles the barking of a ſmall dog. Though a voracious bird, the Eagle can endure the want of ſuſtenance for a long time. A common Eagle, caught in a fox trap, is ſaid to have paſſed five whole weeks without the leaſt food, and did not appear ſenſibly weakened till towards the laſt week, after which a period was put to its exiſtence.

THE GOLDEN EAGLE,

(*Falco Chryſætos*, Linnæus.—*Le grand Aigle*, Buffon.)

Is the largeſt of the genus: It meaſures, from the point of the bill to the extremity of the toes, upwards of three feet; and in breadth, from wing to wing, above eight; and weighs from ſixteen to eighteen pounds. The male is ſmaller, and does not weigh more than twelve pounds. The bill is of a deep blue colour; the cere yellow; the eyes are large, deep ſunk, and covered by a projecting brow; the iris is of a fine bright yellow, and ſparkles with uncommon luſtre. The general colour is deep brown, mixed with tawny on the head and neck; the quills are chocolate, with white ſhafts; the tail is black, ſpotted with aſh colour; the legs are yellow, and feathered down to the toes, which are very ſcaly; the claws are remarkably large; the middle one is two inches in length.—This noble bird is found in various parts of Europe; it abounds moſt in the warmer regions, and has ſeldom been met with farther north than the fifty-fifth degree of latitude. It is known to breed in the mountainous parts of Ireland; it lays three, and ſometimes four eggs, of which it ſeldom happens that more than two are prolific. Mr Pennant ſays there are inſtances, though rare, of their having bred in Snowdon Hills. Mr Wallis, in his Natural Hiſtory of Northumberland, ſays, it formerly had its aery on the higheſt and ſteepeſt part of Cheviot. In the beginning of January, 1735, a very large one was ſhot near Warkworth, which meaſured, from point to point of its wings, eleven feet and a quarter.

THE RINGTAILED EAGLE.

(*Falco Fulvus.* Lin.—*L'Aigle Commun.* Buff.)

This is the common Eagle of Buffon, and, according to that author, includes two varieties, the brown and the black Eagle; they are both of the ſame brown colour, diſtinguiſhed only by a deeper ſhade; and are nearly of the ſame ſize. In both, the upper part of the head and neck is mixed with ruſt colour, and the baſe of the larger feathers marked with white; the bill is of a dark horn colour, the cere of a bright yellow, the iris hazel, and between the bill and the eye there is a naked ſkin of a dirty brown colour; the legs are feathered to the toes, which are yellow, and the claws black; the tail is diſtinguiſhed by a white ring, which covers about two thirds of its length; the remaining part is black.

The Ringtailed Eagle is more numerous and diffuſed than the Golden Eagle, and prefers more northern climates. It is found in France, Germany, Switzerland, Great Britain, and in America as far north as Hudſon's Bay.

THE WHITE-TAILED EAGLE.

GREAT ERNE—CINEREOUS EAGLE.

(*Falco Albiulla*, Lin.—*Le grand Pygargue*, Buff.)

Of this there appear to be three varieties, which differ chiefly in ſize, and conſiſt of the following: the great Erne, or Cinereous Eagle, of Latham and Pennant; the ſmall Erne, or leſſer White-tailed Eagle; and the White-headed Erne, or Bald Eagle. The two firſt are diſtinguiſhed only by their ſize, and the laſt by the whiteneſs of its head and neck.

The white-tailed Eagle is inferior in ſize to the Golden Eagle; the beak, cere, and eyes are of a pale yellow; the ſpace between the beak and the eye is of a blueiſh colour, and thinly covered with hair; the ſides of the head and neck are of a pale aſh colour, mixed with reddiſh brown; the general colour of the plumage is brown, darkeſt on the upper part of the head, neck, and back; the quill feathers are very dark; the breaſt is irregularly marked with white ſpots; the tail is white; the legs, which are of a bright yellow, are feathered a little below the knees; the claws are black.

This bird inhabits all the northern parts of Europe, and is found in Scotland and many parts of Great-Britain; it is equal in ſtrength and vigour to the common Eagle, but more furious; and is ſaid to drive its young ones from the neſt, after having fed them only a very ſhort time. It has commonly two or three young, and builds its neſt upon lofty trees.

THE SEA EAGLE.

(*Falco Offifragus*, Lin.—*L'Orfraie*, Buff.)

This bird is nearly as large as the Golden Eagle, meafuring in length three feet and a half, but its expanded wings do not reach above feven feet,

Its bill is large, much hooked, and of a blueiſh colour; its eye is yellow; a row of ſtrong briſtly feathers hangs down from its under bill next to its throat, from whence it has been termed the bearded Eagle; the top of the head and back part of the neck are dark brown, inclining to black; the feathers on the back are variegated by a lighter brown, with dark edges; the ſcapulars are pale brown, the edges nearly white; the breaſt and belly whitiſh, with irregular ſpots of brown; the tail feathers are dark brown; the outer edges of the exterior feathers whitiſh; the quill feathers and thighs are duſky; the legs and feet yellow; the claws, which are large, and form a compleat ſemicircle, are of a ſhining black. It is found in various parts of Europe and America; it is ſaid to lay only two eggs during the whole year, and frequently produces only one young one; it is however widely diſperſed, and was met with at Botany Iſland by Captain Cook. It lives chiefly on fiſh; its uſual haunts are by the ſea-ſhore; it alſo frequents the borders of large lakes and rivers; and is ſaid to ſee ſo diſtinctly in the dark, as to be able to purſue and catch its prey during the night. The ſtory of the Eagle, brought to the ground after a ſevere conflict with a cat which it had ſeized and taken up into the air with its talons, is very remarkable. Mr Barlow, who was an eye-witneſs of the fact, made a drawing of it, which he afterwards engraved.

THE OSPREY.

BALD BUZZARD, SEA EAGLE, OR FISHING HAWK.

(*Falco Haliætus*, Lin.—*Le Balbuzard*, Buff.)

The length of this bird is two feet; its breadth from tip to tip, above five; its bill is black, with a blue cere, and its eye is yellow; the crown of its head is white, marked with oblong dusky spots; its cheeks, and all the under parts of its body, are white, slightly spotted with brown on its breast; from the corner of each eye a streak of brown extends down the ſides of the neck toward the wing; the upper part of the body is brown; the two middle feathers of the tail are brown, the others are marked on the inner webs with alternate bars of brown and white; the legs are very ſhort and thick, being only two inches and a quarter long, and two inches in circumference; they are of a pale blue colour; the claws black; the outer toe is larger than the inner one, and turns eaſily backward, by which means this bird can more readily ſecure its ſlippery prey.

Buffon obſerves that the Oſprey is the moſt numerous of the large birds of prey, and is ſcattered over the extent of Europe, from Sweden to Greece, and that it is found even in Egypt and Nigritia. Its haunts are on the ſea ſhore, and on the borders of rivers and lakes; its principal food is fiſh; it darts upon its prey with great rapidity, and with undeviating aim. The Italians compare its deſcent upon the water to a piece of lead falling upon that element, and diſtinguiſh it by the name of Auguiſta Piumbina, or the Leaden Eagle. It builds its neſt on the ground, among reeds, and lays three or four eggs, of an elliptical form, rather leſs than thoſe of a hen. The Carolina and Cayenne Oſpreys are varieties of this ſpecies.

THE COMMON BUZZARD.

(*Falco Buteo*, Lin.—*La Buſe*, Buff.)

M. Buffon diſtinguiſhes the Kites and the Buzzards from the Eagles and Hawks by their habits and diſpoſitions, which he compares to thoſe of the Vultures, and places them after thoſe birds. Though poſſeſſed of ſtrength, agility, and weapons to defend themſelves, they are cowardly, inactive, and ſlothful; they will fly before a Sparrow-hawk, and when overtaken will ſuffer themſelves to be beaten, and even brought to the ground without refiftance.

The Buzzard is about twenty inches in length, and in breadth four feet and a half; its bill is of a lead colour; eyes pale yellow; the upper parts of the body are of a dufky brown colour; the wings and tail are marked with bars of a darker hue; the upper parts pale, variegated with a light reddifh brown; the legs are yellow; claws black. This well-known bird is of a fedentary and indolent difpofition; it continues for many hours perched upon a tree or eminence, from whence it darts upon the game that comes within its reach; it feeds on birds, fmall quadrupeds, reptiles, and infects; its neft is conftructed with fmall branches, lined in the infide with wool, and other foft materials; it lays two or three eggs, of a whitifh colour, fpotted with yellow; it feeds and tends its young with great affiduity. Ray affirms, that if the female be killed during the time of incubation, the male Buzzard takes the charge of them, and patiently rears the young till they are able to provide for themfelves. Birds of this fpecies are fubject to greater variations than moft other birds, fcarcely two being alike; fome are entirely white, of others the head only is white, and others again are mottled with brown and white.

We were favoured with one of thefe birds by John Trevelyan, Efq. of Wallington, in the county of Northumberland, by whom it was fhot in the act of devouring its prey, which consisted of a partridge it had just killed: The flesh was entirely separated from the bones, which, with the legs and wings, were afterwards discovered laying at a small distance from the place where it had been shot.

THE HONEY BUZZARD.

(*Falco Apivorus*, Lin.—*La Bondree*, Buff.)

Is as large as the Buzzard, measuring twenty-two inches in length; its wings extend above four feet; its bill is black, and rather longer than that of the Buzzard; its eyes are yellow; its head is large and flat, and of an aſh colour; upper parts of the body dark brown; the under parts white, ſpotted or barred with ruſty brown on the breaſt and belly; tail brown, marked with three broad duſky bars, between each of which are two or three of the ſame colour, but narrower; the legs are ſtout and ſhort, of a dull yellow colour; claws black. This bird builds its neſt ſimilar to that of the Buzzard, and of the ſame materials; its eggs are of an aſh colour, with ſmall brown ſpots: It ſometimes takes poſſeſſion of the neſts of other birds, and feeds its young with waſps and other inſects; it is fond of field mice, frogs, lizards, and inſects: it does not ſoar like the Kite, but flies low from tree to tree, or from buſh to buſh: It is found in all the northern parts of Europe, and in the open parts of Ruſſia and Siberia, but is not ſo common in England as the Buzzard.

Buffon obſerves, that it is frequently caught in the winter, when it is fat and delicious eating.

MOOR BUZZARD.

DUCK HAWK, OR WHITE-HEADED HARPY.

(*Falco Æruginoſus*, Lin.—*Le Buſard*, Buff)

LENGTH above twenty-one inches; the bill is black; cere and eyes yellow; the whole crown of the head is of a yellowiſh white, lightly tinged with brown; the throat is of a light ruſt colour; the reſt of the plumage is of a reddiſh brown, with pale edges; the greater wing coverts tipped with white; the legs are yellow; claws black. Our figure and defcription are taken from a very fine living bird fent us by John Silvertop, Efq. of Minfter-Acres, in the county of Northumberland, which very nearly agreed with that figured in the *Planches Enluminees*. Birds of this kind vary much—in fome the crown and back part of the head are yellow; and in one defcribed by Mr Latham, the whole bird was uniformly of a chocolate brown, with a tinge of ruft colour. It preys on rabbits, young wild ducks, and other water fowl, and likewife feeds on fifh, frogs, reptiles, and even infects: Its haunts are in hedges and bufhes near pools, marfhes, and rivers, that abound with fifh; it builds its neft a little above the furface of the ground, or in hillocks covered with thick herbage; the female lays three or four eggs, of a whitifh colour, irregularly fprinkled with dufky fpots:—Though fmaller, it is more active and bolder than the Common Buzzard; and when purfued, it meets its antagonift, and makes a vigorous defence.

THE KITE.

PUTTOK, FORK-TAILED KYTE, OR GLEAD.

(*Falco Milvus*, Lin.—*Le Milan Royal*, Buff.)

This bird is eaſily diſtinguiſhed from the Buzzard by its forked tail, which is its peculiar and diſtinguiſhing feature: Its length is about two feet; its bill is of a horn colour, furniſhed with briſtles at its baſe; its eyes and cere are yellow; the feathers on the head and neck are long and narrow, of a hoary colour, ſtreaked with brown down the middle of each; the body is of a reddiſh brown colour, the margin of each feather being pale; the quills are dark brown; the legs yellow; and the claws black. It is common in England, and continues with us the whole year: It is found in various parts of Europe, in very northern latitudes, from whence before winter it retires towards Egypt in great numbers; it is ſaid to breed there, and return in April to Europe, where it breeds a ſecond time, contrary to the nature of rapacious birds in general. The female lays two or three eggs of a whitiſh colour, ſpotted with pale yellow, and of a roundiſh form. Though the Kite weighs ſomewhat leſs than three pounds, the extent of its wings is more than five feet; its flight is rapid, and it ſoars very high in the air, frequently beyond the reach of our ſight,—yet at this diſtance it perceives its food diſtinctly, and deſcends upon its prey with irreſiſtible force; its attacks are confined to ſmall animals and birds; it is particularly fond of young chickens, but the fury of their mother is generally ſufficient to drive away the robber.

THE GOSHAWK.

(*Falco Palumbarius*, Lin.—*L'Autour*, Buff.)

THIS bird is ſomewhat longer than the Buzzard, but ſlenderer and more beautiful; its length is one foot ten inches; its bill is blue, tipped with black; cere green; eyes yellow; over each eye there is a whitiſh line; the head and all the upper parts of the body are of a deep brown colour, each ſide of the neck being irregularly marked with white; the breaſt and belly are white, with a number of wavy lines or bars of black; the tail is long, of an aſh colour, and croſſed with four or five duſky bars; the legs are yellow, and the claws black; the wings are much ſhorter than the tail. M. de Buffon, who brought up two young birds of this kind, a male and a female, makes the following obſervations: That the Goſhawk, before it has ſhed its feathers, that is, in its firſt year, is marked on the breaſt and belly with longitudinal brown ſpots; but after it has had two moultings they diſappear, and their place is occupied by tranſverſe bars, which continue during the reſt of its life: He obſerves further, that though the male was much ſmaller than the female, it was fairer and more vicious: The Goſhawk feeds on mice and ſmall birds, and eagerly devours raw fleſh; it plucks the birds very neatly, and tears them into pieces before it eats them, but ſwallows the pieces entire; and frequently diſgorges the hair rolled up in ſmall pellets.

The Goſhawk is found in France and Germany; it is not very common in this country, but is more frequent in Scotland; it is likewiſe common in North America, Ruſſia, and Siberia: In Chineſe Tartary there is a variety which is mottled with brown and yellow. They are ſaid to be uſed by the Emperor of China in his ſporting excurſions, when he is uſually attended by his grand falconer, and a thouſand of inferior rank. Every bird has a ſilver plate faſtened to its foot, with the name of the falconer who had the charge of it, that in caſe it ſhould be loſt it may be reſtored to the proper perſon; but if he ſhould not be found, the bird is delivered to another officer called *the guardian of loſt birds*, who, to make his ſituation known, erects his ſtandard in a conſpicuous place among the army of hunters. In former times the cuſtom of carrying a hawk on the hand was confined to men of high diſtinction, ſo that it was a ſaying among the Welſh, " you may know a gentleman by his hawk, horſe, and greyhound." Even the ladies in thoſe times were partakers of this gallant ſport, and have been repreſented in ſculpture with hawks on their hands. At preſent this noble diverſion is wholly laid aſide in this country; the advanced ſtate of agriculture which every where prevails, and the conſequent improvement and incloſure of lands, would but ill accord with the purſuits of the falconer, who requires a large and extenſive range of country, where he may purſue his game without moleſtation to himſelf, or injury to his neighbour. The expence which attended this ſport was very conſiderable, which confined it to princes and men of the higheſt rank. In the time of James I. Sir Thomas Monſon is ſaid to have given a thouſand pounds for a caſt of hawks. In the reign of Edward III. it was made felony to ſteal a hawk; to take its eggs, even in a perſon's own ground, was puniſhable with impriſonment for a year and a day, together with a fine at the king's pleaſure. Such was the pleaſure our anceſtors took in this royal ſport, and ſuch were the means by which they endeavoured to ſecure it.—Beſides the bird juſt deſcribed, there are many other kinds which were formerly in high eſtimation for the ſports of the field; theſe were principally the Jer-Falcon, the Falcon, the Lanner, the Sacre, the Hobby, the Keſtril, and the Merlin: Theſe are called the long-winged hawks, and are diſtinguiſhed from the Goſhawk, the Sparrowhawk, the Kite, and the Buzzard, which are of ſhorter wing, ſlower in their motions, more indolent, and leſs courageous than the others.

THE SPARROWHAWK.

(*Falco Nisus*, Lin.—*L'Epervier*, Buff.)

The length of the male is twelve inches; that of the female fifteen: Its bill is blue, furnished with bristles at the base, which overhang the nostrils; the colour of the eye is bright orange; the head is flat at the top, and above each eye is a strong bony projection, which seems as if intended to secure it from external injury; from this projection a few scattered spots of white form a faint line running backward towards the neck; the top of the head and all the upper parts of the body are of a dusky brown colour; on the back part of the head there is a faint line of white; the scapulars are marked with two spots of white on each feather; the greater quill feathers and the tail are dusky, with four bars of a darker hue on each; the inner edges of all the quills are marked with two or more large white ſpots; the tips of the tail feathers are white; the breaſt, belly, and under coverts of the wings and thighs are white, beautifully barred with brown; the throat is faintly ſtreaked with brown; the legs and feet are yellow; claws black.

The above deſcription is that of a female; the male differs both in ſize and colour, the upper part of his body being of a dark lead colour, and the bars on his breaſt more numerous. The Sparrow-hawk is a bold and ſpirited bird, and very numerous in various parts of the world, from Ruſſia to the Cape of Good Hope. The female builds her neſt in hollow trees, high rocks, or lofty ruins, ſometimes in old crows' neſts, and generally lays four or five eggs, ſpotted with reddiſh ſpots at the longer end. The Sparrowhawk is obedient and docile, and can be eaſily trained to hunt partridges and quails; it makes great deſtruction among pigeons, young poultry, and ſmall birds of all kinds, which it will attack and carry off in the moſt daring manner.

THE JER-FALCON.

(*Falco Gyrfalco*, Lin.—*Le Gerfaut*, Buff.)

This is a very elegant ſpecies, and equals the Goſhawk in ſize: Its bill is much hooked, and yellow; the iris is duſky; the throat white, as is likewiſe the general colour of the plumage, ſpotted with brown; the breaſt and belly are marked with lines, pointing downwards; the ſpots on the back and wings are larger; the feathers on the thighs are very long, and of a pure white; thoſe of the tail are barred; the legs are of a pale blue, and feathered below the knee. This bird is a native of the cold and dreary climates of the north, being found in Ruſſia, Norway, and Iceland; it is never ſeen in warm, and ſeldom in temperate climates; it is found, but rarely, in Scotland and the Orkneys. Buffon mentions three varieties of the Jer-Falcon; the firſt is brown on all the upper parts of the body, and white ſpotted with brown on the under: This is found in Iceland: The ſecond is very ſimilar to it; and the third is entirely white. Next to the Eagle, it is the moſt formidable, the moſt active, and the moſt intrepid of all voracious birds, and is the deareſt and moſt eſteemed for falconry: It is tranſported from Iceland and Ruſſia into France, Italy, and even into Perſia and Turkey—nor does the heat of theſe climates appear to diminiſh its ſtrength, or blunt its vivacity. It boldly attacks the largeſt of the feathered race; the Stork, the Heron, and the Crane are eaſy victims: It kills hares by darting directly upon them. The female, as in all other birds of prey, is much larger and ſtronger than the male, which is uſed in falconry only to catch the Kite, the Heron, and the Crow.

THE GENTIL-FALCON.

(*Falco Gentilis*, Lin.)

THIS bird is ſomewhat larger than the Goſhawk: Its bill is lead colour; cere and irides yellow; the head and back part of the neck are ruſty ſtreaked with black; the back and wings are brown; ſcapulars tipped with ruſty; the quills duſky; the outer webs barred with black; the lower part of the inner webs marked with white; the tail is long, and marked with alternate bars of black and aſh colour, and tipped with white; the legs are yellow, and the claws black; the wings do not extend farther than half the length of the tail.

Naturaliſts have enumerated a great variety of Falcons: and in order to ſwell the liſt, they have introduced the ſame bird at different periods of its life; and have, not unfrequently, ſubſtituted accidental differences of climate as conſtituting permanent varieties; ſo that, as Buffon obſerves with his uſual acuteneſs, one would be apt to imagine that there were as many varieties of the Falcon as of the Pigeon, the Hen, and other domeftic birds. In this way new fpecies have been introduced, and varieties multiplied without end: An over-anxious defire of noting all the minute differences exifting in this part of the works of nature has fometimes led the too curious inquirer into unneceffary diftinctions, and has been the means of introducing confufion and irregularity into the fyftems of ornithologifts. Our countryman, Latham, makes twelve varieties of the common Falcon, of which one is a young Falcon, or yearling—another is the Haggard, or old Falcon—whilft others differ only in fome uneffential point, arifing from age, fex, or climate. Buffon, however, reduces the whole to two kinds—the Gentil, which he fuppofes to be the fame with the common Falcon, differing only in feafon; and the Peregrine, or Paffenger Falcon. This laft is rarely met with in Britain, and confequently is but little known with us: It is about the fize of the common Falcon; its bill is blue, black at the point; cere and irides yellow; the upper parts of the body are elegantly marked with bars of blue and black; the breaft is of a yellowifh white, marked with a few fmall dufky lines; the belly, thighs, and vent of a greyifh white, croffed with dufky bands; the quills are dufky, fpotted with white; the tail is finely barred with blue and black; the legs are yellow; the claws black.

THE LANNER.

(*Falco Lanarius*, Lin.—*Le Lanier*, Buff.)

This bird is ſomewhat leſs than the Buzzard: its bill is blue; cere inclining to green; eyes yellow; the feathers on the upper parts of the body are brown, with pale edges; above each eye there is a white line, which runs towards the hind part of the head, and beneath it is a black ſtreak pointing downwards towards the neck; the throat is white; the breaſt of a dull yellow, marked with brown ſpots; thighs and vent the ſame; the quill feathers are duſky, marked on the inner webs with oval ſpots, of a ruſt colour; the tail is ſpotted in the ſame manner; the legs are ſhort and ſtrong, and of a blueiſh colour. The Lanner is not common in England; it breeds in Ireland, and is found in various parts of Europe: It derives its name from its mode of tearing its prey into ſmall pieces with its bill.

THE HEN-HARRIER.

DOVE-COLOURED FALCON, OR BLUE HAWK.

(*Falco Cyaneus*, Lin.—*L'Oiſeau St. Martin*, Buff.)

The length ſeventeen inches; breadth, from tip to tip, ſomewhat more than three feet; the bill is black, and covered at the baſe with long briſtly feathers; the cere, irides, and edges of the eye-lids are yellow; the upper parts of the body are of a blueiſh grey colour, mixed with light tinges of ruſty; the breaſt and under coverts of the wings are white, the former marked with ruſty coloured ſtreaks, the latter with bars of the ſame colour; the greater quills are black, the ſecondaries and leſſer quills aſh-coloured; on the latter, in ſome birds, a ſpot of black in the middle of each feather forms a bar acroſs the wing; the two middle feathers of the tail are grey, the three next are marked on their inner webs with duſky bars, the two outermoſt are marked with alternate bars of white and ruſt colour; the legs are long and ſlender, and of a yellow colour. Theſe birds vary much; of ſeveral which we have been favoured with, from John Silvertop, Eſq. ſome were perfectly white on the under parts, and of a larger ſize than common:—We ſuppoſe the difference ariſes from the age of the bird.*

The Hen-harrier feeds on birds, lizards, and other reptiles; it breeds annually on Cheviot, and on the ſhady precipices under the Roman wall by Crag-lake;† it flies low, ſkimming along the ſurface of the ground in ſearch of its prey: The female makes her neſt on the ground, and lays four eggs of a reddiſh colour, with a few white ſpots.

* It has been ſuppoſed that this and the following are male and female; but the repeated inſtances of Hen-harriers of both ſexes having been ſeen, leaves it beyond all doubt, that they conſtitute two diſtinct ſpecies.

† Wallis's Natural Hiſtory of Northumberland.

THE RINGTAIL.

(*Falco Pygargus*, Lin.—*Soubuſe*, Buff.)

Its length is twenty inches; breadth three feet nine; its bill is black; cere and irides yellow; the upper part of the body is duſky; the breaſt, belly, and thighs are of a yellowiſh brown, marked with oblong duſky ſpots; the rump white; from the back part of the head behind the eyes to the throat there is a line of whitiſh coloured feathers, forming a collar or wreath; under each eye there is a white ſpot; the tail is long, and marked with alternate brown and duſky bars; the legs are yellow; claws black.

THE KESTREL.

STONEGALL, STANNEL HAWK, OR WINDHOVER.

(*Falco Tinnunculus*, Lin.—*La Creſſerelle*, Buff.)

The male of this ſpecies differs ſo much from the female, and is ſo rarely ſeen, that we have given a figure of it from one we had in our poſſeſſion. Its length is fourteen inches; breadth two feet three inches; its bill is blue; cere and eyelids yellow; eyes black; the forehead dull yellow; the top of the head, back part of the neck, and ſides, as far as the points of the wings, are of a lead colour, faintly ſtreaked with black; the cheeks are paler; from the corner of the mouth on each ſide there is a dark ſtreak pointing downwards; the back and coverts of the wings are of a bright vinous colour, ſpotted with black; quill feathers duſky, with light edges; all the under part of the body is of a pale ruſt colour, ſtreaked and ſpotted with black; thighs plain; the tail feathers are of a fine blue grey, with black ſhafts; towards the end there is a broad black bar both on the upper and under ſides; the tips are white; the legs are yellow, and the claws black.

THE FEMALE KESTREL.

This beautiful bird is diſtinguiſhed from every other Hawk by its variegated plumage; its bill is blue; cere and feet yellow; eyes dark coloured, ſurrounded with a yellow ſkin; its head is ruſt coloured, ſtreaked with black; behind each eye there is a bright ſpot; the back and wing coverts are elegantly marked with numerous undulated bars of black; the breaſt, belly, and thighs are of a pale reddifh colour, with dufky ftreaks pointing downwards; vent plain; the tail is marked by a pretty broad black bar near the end, a number of fmaller ones, of the fame colour, occupying the remaining part; the tip is pale.

The Keftrel is widely diffufed throughout Europe, and is found in the more temperate parts of North America: It is a handfome bird, its fight is acute, and its flight eafy and graceful: It breeds in the hollows of trees, and in the holes of rocks, towers, and ruined buildings; it lays four or five eggs, of a pale reddifh colour: Its food confifts of fmall birds, field mice, and reptiles: After it has fecured its prey, it plucks the feathers very dexteroufly from the birds, but fwallows the mice entire, and difcharges the hair at the bill in the form of round balls. This bird is frequently feen hovering in the air, and fanning with its wings by a gentle motion, or wheeling flowly round, at the fame time watching for its prey, on which it fhoots like an arrow. It was formerly ufed in Great Britain for catching fmall birds and young Partridges.

THE HOBBY.

(*Falco Subbuteo*, Lin.—*Le Hobreau*, Buff.)

The length of the male is twelve inches; breadth about two feet; the bill is blue; cere and orbits of the eyes yellow, the irides orange; over each eye there is a light coloured ſtreak; the top of the head, coverts of the wings, and back, are of a dark brown, in ſome edged with ruſt colour; the hind part of the neck is marked with two pale yellow ſpots; a black mark extends from behind each eye, forming almoſt a creſcent, and extending downwards on the neck; the breaſt and belly are pale, marked with duſky ſtreaks; the thighs ruſty with long duſky ſtreaks; the wings brown; the two middle feathers of the tail are of a deep dove colour, the others are barred with ruſty, and tipt with white. The female is much larger, and the ſpots on her breaſt more conſpicuous than thoſe of the male; the legs and feet are yellow.

The Hobby breeds with us, but is ſaid to emigrate in October. It was formerly uſed in falconry, chiefly for Larks and other ſmall birds. The mode of catching them was ſingular; when the Hawk was caſt off, the Larks, fixed to the ground through fear, became an eaſy prey to the fowler, by drawing a net over them. Buffon ſays that it was uſed in hunting Partridges and Quails.

THE MERLIN.

(*Falco Æſalon*, Lin.—*L'Emerillon*, Buff.)

The Merlin is the ſmalleſt of all the Hawk kind, ſcarcely exceeding the ſize of a Blackbird: Its bill is blue; cere yellow; irides very dark; the head is ruſt colour, ſtreaked with black; back and wings of a dark blueiſh aſh colour, ſtreaked and ſpotted with ruſt colour; quill feathers dark, marked with reddiſh ſpots; the breaſt and belly are of a yellowiſh white, with ſtreaks of brown pointing downwards; the tail is long, and marked with alternate duſky and reddiſh bars; the wings, when cloſed, do not reach quite to the end of the tail; the legs are yellow; claws black.

The Merlin, though ſmall, is not inferior in courage to any of the Falcon tribe. It was uſed for taking Larks, Partridges, and Quails, which it would frequently kill by one blow, ſtriking them on the breaſt, head, and neck. Buffon obſerves that this bird differs from the Falcons, and all the rapacious kind, in the male and female being of the ſame ſize. The Merlin does not breed here, but viſits us in October; it flies low, and with great celerity and eaſe; it preys on ſmall birds, and breeds in woods, laying five or ſix eggs.

OF THE OWL.

THE Owl is diſtinguiſhed among birds of the rapacious kind by peculiar and ſtriking characters: Its outward appearance is not more ſingular than are its habits and diſpoſitions; unable to bear the brighter light of the ſun, the Owl retires to ſome obſcure retreat, where it paſſes the day in ſilence and obſcurity, but at the approach of evening, when all nature is deſirous of repoſe, and the ſmaller animals, which are its principal food, are ſeeking their neſtling places, the Owl comes forth from its lurking holes in queſt of its prey. Its eyes are admirably adapted for this purpoſe, being ſo formed as to diſtinguiſh objects with greater facility in the duſk than in broad day-light: Its flight is rapid and ſilent during its nocturnal excurſions, and it is then known only by its frightful and reiterated cries, with which it interrupts the ſilence of the night. During the day, the Owl is ſeldom ſeen; but if forced from his retreat, his flight is broken and interrupted, and he is ſometimes attended by numbers of ſmall birds of various kinds, who, ſeeing his embarraſſment, purſue him with inceſſant cries, and torment him with their movements; the Jay, the Thruſh, the Blackbird, the Redbreaſt, and the Titmouſe all aſſemble to hurry and perplex him. During all this, the Owl remains perched upon the branches of a tree, and anſwers them only with aukward and inſignificant geſtures, turning its head, its eyes, and its body with all the appearance of mockery and affectation. All the ſpecies of Owls, however, are not alike dazzled and confuſed with the light of the ſun, ſome of them being able to fly and ſee diſtinctly in open day.

Nocturnal birds of prey are generally divided into two kinds—thoſe which have horns or ears, and thoſe which are earleſs or without horns; theſe horns conſiſt of ſmall tufts of feathers ſtanding up like ears on each ſide of the head, which may be erected or depreſſed at the pleaſure of the animal; and in all probability are of uſe in directing the organs of hearing, which are very large, to their proper object. Both kinds agree in having their eyes ſo formed as to be able to purſue their prey with much leſs light than other birds. The general character of the Owl is as follows: The eyes are large, and are ſurrounded with a radiated circle of feathers, of which the eye itſelf is the center; the beak and talons are ſtrong and crooked; the body very ſhort, but thick, and well covered with a coat of the ſofteſt and moſt delicate plumage; the external edges of the outer quill feathers in general are ſerrated or finely toothed, which adds greatly to the ſmoothneſs and ſilence of its flight.

We ſhall now proceed to mention thoſe particular ſpecies which are found in this country, and ſhall begin with the largeſt of them.

THE GREAT-EARED OWL.

(*Strix Bubo*, Lin.—*Le grand Duc*, Buff)

This bird is not much inferior in ſize to an Eagle: Its head is very large, and is adorned with two tufts, more than two inches long, which ſtand juſt above each eye; its bill is ſtrong, and much hooked; its eyes large, and of a bright yellow; the whole plumage is of a ruſty brown, finely variegated with black and yellow lines, ſpots, and ſpecks; its belly is ribbed with bars of a brown colour confuſedly intermixed; its tail ſhort, marked with duſky bars; its legs are ſtrong, and covered to the claws with a thick cloſe down, of a ruſt colour; its claws are large, much hooked, and of a duſky colour: Its neſt is large, being nearly three feet in diameter; it is compoſed of ſticks bound together by fibrous roots, and lined with leaves; it generally lays two eggs, ſomewhat larger than thoſe of a Hen, and variegated like the bird itſelf; the young ones are very voracious, and are well ſupplied with various kinds of food by the parents. This bird has been found, though rarely, in Great Britain; it builds its neſt in the caverns of rocks, in mountainous and almoſt inacceſſible places, and is ſeldom ſeen in the plain, or perched on trees; it feeds on young hares, rabbits, rats, mice, and reptiles of various kinds.

THE LONG-EARED OWL.

HORN OWL.

(*Strix Otus*, Lin.—*Le Hibou*, Buff.)

Its length is fourteen inches; breadth ſomewhat more than three feet: Its bill is black; irides of a bright yellow; the radiated circle round each eye is of a light cream colour, in ſome parts tinged with red; between the bill and the eye there is a circular ſtreak, of a dark brown colour; another circle of a dark ruſty brown entirely ſurrounds the face; its horns or ears conſiſt of ſix feathers cloſe-ly laid together, of a dark brown colour, tipped and edged with yellow; the upper part of the body is beautifully penciled with fine ſtreaks of white, ruſty, and brown: the breaſt and neck are yellow, finely marked with duſky ſtreaks, pointing downwards; the belly, thighs, and vent feathers are of a light cream colour: upon each wing there are four or five large white ſpots; the quill and tail feathers are marked with duſky and reddiſh bars; the legs are feathered down to the claws, which are very ſharp; the outer claw is moveable, and may be turned backwards.

This bird is common in various parts of Europe, as well as in this country; its uſual haunts are in old ruined buildings, in rocks, and in hollow trees. M. Buffon obſerves that it ſeldom conſtructs a neſt of its own, but not unfrequently occupies that of the Magpie; it lays four or five eggs; the young are at firſt white, but acquire their natural colour in about fifteen days.

THE SHORT-EARED OWL.

(*Strix Brachyotos*, Phil. Tranſ. vol. 62, p. 384.)

Length fourteen inches; breadth three feet: The head is ſmall, and Hawk-like; bill duſky; the eyes are of a bright yellow, which, when the pupil is contracted, ſhine like gold; the circle round each eye is of a dirty white, with dark ſtreaks pointing outwards; immediately round the eye there is a circle of black; the two horns or ears, in thoſe we have examined, conſiſt of not more than three feathers, of a pale brown or tawny colour, with a dark ſtreak in the middle of each; the whole upper part of the body is variouſly marked with dark brown and tawny, the feathers being moſtly edged with the latter; the breaſt and belly are of a pale yellow, marked with dark longitudinal ſtreaks, which are moſt numerous on the breaſt; the legs and feet are covered with feathers of a pale yellow colour; the claws are much hooked, and black; the wings are long, and extend beyond the tail; the quills are marked with alternate bars of a duſky and pale brown; the tail is likewiſe marked with bars of the ſame colour, the middle feathers of which are diſtinguiſhed by a dark ſpot in the centre of the yellow ſpace; the tip is white. Of ſeveral of theſe birds, both male and female, which we have been favoured with by our friends, we have obſerved that both had the upright tufts or ears: In one of theſe, which was alive in our poſſeſſion, they were very conſpicuous, and appeared more erect while the bird remained undiſturbed; but when frightened, were ſcarcely to be ſeen;—in the dead birds they were hardly diſcernible.

Mr Pennant ſeems to be the firſt deſcriber of this rare and beautiful bird, which he ſuppoſes to be a bird of paſſage, as it only viſits us the latter end of the year, and returns in the ſpring to the places of its ſummer reſidence. It is found chiefly in wooded or mountainous countries: Its food is principally field mice, of which it is very fond.

THE FEMALE HORNED OWL.

This bird was ſomewhat larger than the former; the colours and marks were the ſame, but much darker, and the ſpots on the breaſt larger and more numerous; the ears were not diſcernible; being a dead bird, and having not ſeen any other at the time it was in our poſſeſſion, we ſuppoſed it to be a diſtinct kind—but having ſince ſeen ſeveral, both males and females, we are convinced of our miſtake.

THE WHITE OWL.

BARN OWL, CHURCH OWL, GILLIHOWLET, OR SCREECH-OWL.

(*Strix Flammea*, Lin.—*L'Effraie, ou la Frefaie*, Buff.)

Length fourteen inches: Bill pale horn colour; eyes dark; the radiated circle round the eye is compofed of feathers of the moft delicate foftnefs, and perfectly white; the head, back, and wings, are of a pale chefnut, beautifully powdered with very fine grey and brown fpots, intermixed with white; the breaſt, belly, and thighs are white; on the former are a few dark ſpots; the legs are feathered down to the toes, which are covered with ſhort hairs; the wings extend beyond the tail, which is ſhort, and marked with alternate bars of duſky and white; the claws are white. Birds of this kind vary conſiderably; of ſeveral which we have had in our poſſeſſion, the differences were very conſpicuous, the colours being more or leſs faint according to the age of the bird; the breaſt in ſome was white, without ſpots—in others pale yellow. The White Owl is well known, and is often ſeen in the moſt populous towns, frequenting churches, old houſes, maltings, and other uninhabited buildings, where it continues during the day, and leaves its haunts in the evening in queſt of its prey: Its flight is accompanied with loud and frightful cries, from whence it is denominated the Screech Owl; during its repoſe it makes a blowing noiſe, reſembling the ſnoring of a man. It makes no neſt, but depoſits its eggs in the holes of walls, and lays five or ſix, of a whitiſh colour. It feeds on mice and ſmall birds, which it ſwallows whole, and afterwards emits the bones, feathers, and other indigeſtible parts, at its mouth, in the form of ſmall round cakes, which are often found in the empty buildings which it frequents.

THE TAWNY OWL.

COMMON BROWN IVY OWL, OR HOWLET.

(*Strix Stridula*, Lin.—*Le Chathuant*, Buff.)

Is about the ſize of the laſt: Its bill is white; eyes dark blue; the radiated feathers round the eyes are white, finely ſtreaked with brown; the head, neck, back, wing coverts, and ſcapulars are of a tawny brown colour, finely powdered and ſpotted with dark brown and black; on the wing coverts and ſcapulars are ſeveral large white ſpots, regularly placed, ſo as to form three rows; the quill feathers are marked with alternate bars of light and dark brown; the breaſt and belly are of a pale yellow, marked with narrow dark ſtreaks pointing downwards, and croſſed with others of the ſame colour; the legs are feathered down to the toes; the claws are large, much hooked, and white. This ſpecies is found in various parts of Europe; it frequents woods, and builds its neſt in the hollows of trees.

THE LITTLE OWL.

(*Strix Paſſerina*, Lin.—*La Cheveche ou petite Chouette*, Buff.)

THIS is the ſmalleſt of the Owl kind, being not larger than a Blackbird: Its bill is brown at the baſe, and of a yellow colour at the tip; eyes pale yellow; the circular feathers on the face are white, tipped with black; the upper part of the body is of an olive brown colour; the top of the head and wing coverts are ſpotted with white; the breaſt and belly white, ſpotted with brown; the feathers of the tail are barred with ruſt colour and brown, and tipped with white; the legs are covered with down of a ruſty colour, mixed with white; the toes and claws are of a browniſh colour. It frequents rocks, caverns, and ruined buildings, and builds its neſt, which is conſtructed in the rudeſt manner, in the moſt retired places: It lays five eggs, ſpotted with white and yellow. It ſees better in the day-time than other nocturnal birds, and gives chace to ſwallows and other ſmall birds on the wing; it likewiſe feeds on mice, which it tears in pieces with its bill and claws, and ſwallows them by morſels: It is ſaid to pluck the birds which it kills, before it eats them, in which it differs from all the other Owls. It is rarely met with in England: It is ſometimes found in Yorkſhire, Flintſhire, and in the neighbourhood of London.

OF THE SHRIKE.

THE laſt claſs we ſhall mention of birds of the rapacious kind is that of the Shrike, which, as M. Buffon obſerves, though they are ſmall and of a delicate form, yet their courage, their appetite for blood, and their hooked bill, entitle them to be ranked with the boldeſt and the moſt ſanguinary of the rapacious tribe. This genus has been variouſly placed in the ſyſtems of naturaliſts; ſometimes it has been claſſed with the Falcons, ſometimes with the Pies, and has even been ranked with the harmleſs and inoffenſive tribes of the Paſſerine kind, to which indeed, in outward appearance at leaſt, it bears no ſmall reſemblance. Conformable, however, to the lateſt arrangements, we have placed it in the rear of thoſe birds which live by rapine and plunder; and, like moſt of the connecting links in the great chain of nature, it will be found to poſſeſs a middle quality, partaking of thoſe which are placed on each ſide of it, and making thereby an eaſy tranſition from the one to the other.

The Shrike genus is diſtinguiſhed by the following characteriſtics: The bill is ſtrong, ſtraight at the baſe, and hooked or bent towards the end; the upper mandible is notched near the tip, and the baſe is furniſhed with briſtles; it has no cere; the tongue is divided at the end; the outer toe is connected to the middle one as far as the firſt joint. To theſe exterior marks we may add, that it poſſeſſes the moſt undaunted courage, and will attack birds much larger and ſtronger than itſelf, ſuch as the Crow, the Magpie, and moſt of the ſmaller kinds of Hawks; if any of theſe ſhould fly near the place of its retreat, the Shrike darts upon them with loud cries, attacks the invader, and drives it from its neſt. The parent birds will ſometimes join on ſuch occaſions; and there are few birds that will venture to abide the conteſt. Shrikes will chace all the ſmall birds upon the wing, and ſometimes will venture to attack Partridges, and even young hares. Thruſhes, Blackbirds, and ſuch like, are their common prey; they fix on them with their talons, ſplit the ſkull with their bill, and feed on them at leiſure.

There are three kinds found in this kingdom, of which the following is the largeſt.

GREAT ASH-COLOURED SHRIKE.

MURDERING PIE, OR GREAT BUTCHER BIRD.

(*Lanius, excubitor*, Lin.—*La Pie-griesche grise*, Buff.)

The length about ten inches: Its bill is black, and furnished with bristles at the base; the upper parts of its plumage are of a pale blue ash colour; the under parts white; a black stripe passes through each eye; the greater quills are black, with a large white spot at the base, forming a bar of that colour across the wing; the lesser quills are white at the top; the scapulars are white; the two middle feathers of the tail are black; the next on each side are white at the ends, which gradually increases to the outermost, which are nearly white; the whole, when the tail is spread, forms a large oval spot of black; the legs are black. The female differs little from the male; it lays ſix eggs, of a dull o-live green, ſpotted at the end with black. Our fi-gure and deſcription were taken from a very fine ſpecimen, ſent us by Lieut. H. F. Gibſon, of the 4th dragoons: It is rarely found in the cultivated parts of the country, preferring the mountainous wilds, among furz and thorny thickets, for its reſi-dence. M. Buffon ſays it is common in France, where it continues all the year: It is met with likewiſe in Ruſſia, and various parts of Europe; it preys on ſmall birds, which it ſeizes by the throat, and, after ſtrangling, fixes them on a ſharp thorn, and tears them in pieces with its bill. Mr Pen-nant obſerves, that, when kept in the cage, it ſticks its food againſt the wires before it will eat it. It is ſaid to imitate the notes of the ſmaller ſinging birds, thereby drawing them near its haunts, in or-der more ſecurely to ſeize them.

THE RED-BACKED SHRIKE.

LESSER BUTCHER BIRD, OR FLUSHER.

(*Lanius Collurio*, Lin.—*L'Ecorcheur*, Buff.)

Is ſomewhat leſs than the laſt, being little more than ſeven inches long: Its bill is black; irides hazel; the head and lower part of the back are of a light grey colour; the upper part of the back and coverts of the wings are of a bright ruſty red; the breaſt, belly, and ſides of a fine pale roſe or bloom colour; the throat is white; a ſtroke of black paſſes from the bill through each eye; the two middle feathers of the tail are black, the others are white at the baſe; the quills are of a brown colour; the legs black.

The female is ſomewhat larger than the male; the head is of a ruſt colour, mixed with grey; the breaſt, belly, and ſides of a dirty white; the tail deep brown; the exterior web of the outer feathers white. It builds its neſt in hedges or low buſhes, and lays ſix white eggs, marked with a reddiſh brown circle towards the larger end. Its manners are ſimilar to the laſt: It frequently preys on young birds, which it takes in the neſt; it likewiſe feeds on graſshoppers, beetles, and other inſects. Like the laſt, it imitates the notes of other birds, in order the more ſurely to decoy them.—When ſitting on the neſt, the female ſoon diſcovers herſelf at the approach of any perſon, by her loud and violent outcries.

THE WOODCHAT,

(*La Pie-Grieſche Rouſſe*, Buff.)

Is ſaid to equal the laſt in point of ſize: Its bill is horn-coloured, feathers round the baſe whitiſh; head and hind part of the neck bright bay; from the baſe of the bill a black ſtreak paſſes through each eye, inclining downwards on the neck; back duſky, under parts of a yellowiſh white; quills black, near the bottom of each a white ſpot; the two middle feathers of the tail are black, the outer edges and tips of the others are white; the legs black. The deſcription of this bird ſeems to have been taken from a drawing by Mr Edwards, in the

Sloanian Museum, and is not unlike the least Butcher Bird of that celebrated naturalist, which it resembles in size and in the distribution of its colours. M. Buffon supposes it may be a variety of the Red-backed Shrike, as they both depart in September, and return at the same time in the spring; the manners of both are said to be the same, and the difference of colours not very material: The female is somewhat different; the upper parts of the plumage being of a reddish colour, transversely streaked with brown; the under parts of a dirty white, marked in the same manner with brown; the tail is of a reddish brown, with a dusky mark near the end, tipt with red.

BIRDS OF THE PIE KIND

Constitute the next order in the arrangement of the feathered part of the creation; they confiſt of a numerous and irregular tribe, widely differing from each other in their habits, appetites, and manners, as well as in their form, ſize, and appearance. In general they are noiſy, reſtleſs, and loquacious, and of all other kinds contribute the leaſt towards ſupplying the neceſſities or the pleaſures of man. At the head of theſe we ſhall place the Crow and its affinities, well known, by its ſooty plumage and croaking note, from every other tribe of the feathered race. Birds of this kind are found in every part of the known world, from Greenland to the Cape of Good Hope; and though generally diſliked for their diſguſting and indiſcriminating voracity, yet in many reſpects they may be ſaid to be of ſingular benefit to mankind, not only by devouring putrid fleſh, but principally by deſtroying great quantities of noxious inſects, worms, and reptiles. Rooks, in particular, are fond of the erucæ of the hedge-chafer, or cheſnut brown beetle, for which they ſearch with indefatigable pains.*

* Theſe inſects appear in hot weather, in formidable numbers, diſrobing the fields and trees of their verdure, bloſſoms, and fruit, ſpreading deſolation and deſtruction wherever they go.— They appeared in great numbers in Ireland during a hot ſummer, and committed great ravages. In the year 1747 whole

They are often accused of feeding on the corn just after it has been sown, and various contrivances have been made both to kill and frighten them away; but, in our estimation, the advantages derived from the destruction which they make among grubs, earth-worms, and noxious insects of various kinds, will greatly overpay the injury done to the future harvest by the small quantity of corn they may destroy in searching after their favorite food. In general they are sagacious, active, and faithful to each other: They live in pairs, and their mutual attachment is constant. They are a clamourous race, mostly build in trees, and form a kind of society, in which there appears something like a regular government; a centinel watches for the general safety, and give notice on the appearance of danger. On the approach of an enemy or a stranger they act in concert, and drive him away with repeated attacks. On these occasions they are as bold as they are artful and cunning, in avoiding the smallest appearance of real danger; of this the disappointed fowler has frequently occasion to take notice, on seeing the birds fly away before he can draw near enough to shoot them; from this circumstance it has been said that they discover their

meadows and corn-fields were destroyed by them in Suffolk.—The decrease of rookeries in that county was thought to be the occasion of it. The many rookeries with us is in some measure the reason why we have so few of these destructive animals.—*Wallis's History of Northumberland.*

danger by the quicknefs of their fcent, which enables them to provide for their fafety in time; but of this we have our doubts, and would rather afcribe it to the quicknefs of their fight, by which they difcover the motions of the fportfman.

The general characters of this kind are well known, and are chiefly as follow:—The bill is ftrong, and has a flight curvature along the top of the upper mandible; the edges are thin, and fharp or cultrated; in many of the fpecies there is a fmall notch near the tip; the noftrils are covered with briftles; tongue divided at the end; three toes forward, one behind, the middle toe connected to the outer as far as the firft joint.

THE RAVEN.

GREAT CORBIE CROW.

(*Corvus Corax*, Lin.—*Le Corbeau*, Buff.)

Is the largeſt of this kind; its length is above two feet, breadth four: Its bill is ſtrong, and very thick at the baſe; it meaſures ſomewhat more than two inches and a half in length, and is covered with ſtrong hairs or briſtles, which extend above half its length, covering the noſtrils; the general colour of the upper parts is of a fine gloſſy black, reflecting a blue tint in particular lights; the under parts are duller, and of a duſky hue.

The Raven is well known in all parts of the

world, and, in times of ignorance and ſuperſtition, was conſidered as ominous, foretelling future events by its horrid croakings, and announcing impending calamities: In theſe times the Raven was conſidered as a bird of vaſt importance, and the various changes and modulations of its voice were ſtudied with the moſt careful attention, and were made uſe of by artful and deſigning men to miſlead the ignorant and unwary. It is a very long-lived bird, and is ſuppoſed ſometimes to live a century or more. It is fond of carrion, which it ſcents at a great diſtance; it is ſaid that it will deſtroy rabbits, young ducks, and chickens; it has been known to ſeize on young lambs which have been dropped in a weak ſtate, and pick out their eyes while yet alive: It will ſuck the eggs of other birds; it feeds alſo on earth-worms, reptiles, and even ſhell-fiſh, when urged by hunger. It may be rendered very tame and familiar, and has been frequently taught to pronounce a variety of words: It is a crafty bird, and will frequently pick up things of value, ſuch as rings, money, &c, and carry them to its hiding-place. It makes its neſt early in the ſpring, and builds in trees and the holes of rocks, laying five or ſix eggs, of a pale blueiſh green colour, ſpotted with brown. The female ſits about twenty days, and is conſtantly attended by the male, who not only provides her with abundance of food, but relieves her by turns, and takes her place in the neſt.

The natives of Greenland eat the fleſh, and make a covering for themſelves with the ſkins of theſe birds, which they wear next their bodies.

THE CARRION CROW.

MIDDEN CROW, OR BLACK-NEBBED CROW.

(*Corvus Corone*, Lin.—*La Corneille*, Buff.)

Is leſs than the Raven, but ſimilar to it in its habits, colour, and external appearance: It is about eighteen inches in length; its breadth above two feet. Birds of this kind are more numerous and as widely ſpread as the Raven; they live moſtly in woods, and build their neſts on trees; the female lays five or ſix eggs, much like thoſe of a Raven. They feed on putrid fleſh of all ſorts; likewiſe on eggs, worms, inſects, and various ſorts of grain. They live together in pairs, and remain with us during the whole year.

THE HOODED CROW.

ROYSTON CROW.

(*Corvus Cornix*, Lin.—*Le Corneille Mantelée*, Buff.)

Is ſomewhat larger and more bulky than the Rook, meaſuring twenty-two inches in length, and twenty-three in breadth: Its bill is black, and two inches long; the head, forepart of the neck wings, and tail are black; the back and all the under parts are of a pale aſh colour; the legs black. This bird arrives with the Woodcock, and on its firſt coming frequents the ſhores of rivers, and departs in the ſpring to breed in other countries, but it is ſaid that they do not entirely leave us, as they have been feen, during the fummer months, in the northern parts of our ifland, where they frequent the mountainous parts of the country, and breed in the pines. In more northern parts it continues the whole year, and fubfifts on fea-worms, fhell-fifh, and other marine productions. With us it is feen to mix with the Rook, and feeds in the fame manner with it. During the breeding feafon they live in pairs, lay fix eggs, and are faid to be much attached to their offspring.

THE ROOK.

(*Corvus Frugilegus*, Lin.—*Le Freux*, Buff.)

This bird is about the ſize of the Carrion Crow, and, excepting its more gloſſy plumage, very much reſembles it: The baſe of the bill, noſtrils, and even round the eyes are covered with a rough ſcabrous ſkin, in which it differs from all the reſt, occaſioned, it is ſaid, by thruſting its bill into the earth in ſearch of worms; but as the ſame appearance has been obſerved in ſuch as have been brought up tame and unaccuſtomed to that mode of ſubſiſtence, we are inclined to conſider it as an original peculiarity. We have already had occasion to observe that they are useful in preventing a too great increase of that destructive insect the chafer or dor-beetle, and by that means make large recompense for the depredations they may occasionally make on the corn fields. Rooks are gregarious, and fly in immense flocks at morning and evening to and from their roosting places in quest of food. During the breeding time they live together in large societies, and build their nests on trees close to each other, frequently in the midst of large and populous towns. These rookeries, however, are often the scenes of bitter contests, the new comers being frequently driven away by the old inhabitants, their half-built nests torn in pieces, and the unfortunate couple forced to begin their work anew in some more undisturbed situation;—of this we had a remarkable instance in Newcastle. In the year 1783 a pair of Rooks, after an unsuccessful attempt to establish themselves in a rookery at no great distance from the Exchange, were compelled to abandon the attempt, and take refuge on the spire of that building, and altho' constantly interrupted by other Rooks, they built their nest on the top of the vane, and brought forth their young, undisturbed by the noise of the populace below them; the nest and its inhabitants were consequently turned about by every change of the wind. They returned and built their nest every year on the same place till 1793, after which the spire was taken down.

THE JACK-DAW.

(*Corvus Monedula*, Lin.—*Le Choucas*, Buff.)

This bird is confiderably lefs than the Rook, being only thirteen inches in length: Its bill is black; eyes white; the hind part of the head and neck are of a hoary grey colour; the reft of the plumage is of a fine gloffy black above, beneath it has a dufky hue; the legs are black.

The Daw is very common in England, and remains with us the whole year: In other countries, as in France and various parts of Germany, it is migratory. They frequent churches, old towers, and ruins, in great flocks, where they build their nefts: The female lays five or fix eggs, paler than thofe of the Crow, and fmaller; they rarely build in trees:—In Hampſhire they ſometimes breed in the rabbit burrows.* They are eaſily tamed, and may be taught to pronounce ſeveral words; they will conceal part of their food, and with it ſmall pieces of money, or toys. They feed on inſects, grain, fruit, and ſmall pieces of meat, and are ſaid to be fond of Partridge eggs. There is a variety of the Daw found in Switzerland, having a white collar round its neck. In Norway and other cold countries they have been ſeen perfectly white.

* White's Natural Hiſtory of Selborne.

THE MAGPIE.

PIANET.

(*Corvus Pica*, Lin.—*La Pie*, Buff.)

Its length is about eighteen inches: Bill ſtrong and black; eyes hazel; the head, neck, and breaſt are of a deep black, which is finely contraſted with the ſnowy whiteneſs of the breaſt and ſcapulars; the neck feathers are very long, extending down the back, leaving only a ſmall ſpace, of a greyiſh aſh colour, between them and the tail coverts, which are black; the plumage in general is gloſſed with green, purple, and blue, which catch the eye in different lights; its tail is very long, and wedge-ſhaped; the under tail coverts, thighs, and legs are black; on the throat and part of the neck there is a kind of feathers, mixed with the others, reſembling ſtrong whitiſh hairs. This beautiful bird is every where very common in England; it is likewiſe found in various parts of the Continent, but not ſo far north as Lapland, nor farther ſouth than Italy: It is met with in America, but not commonly, and is migratory there: It feeds, like the Crow, on almoſt every thing animal as well as vegetable. The female builds her neſt with great art, leaving a hole in the ſide for her admittance, and covering the whole upper part with a texture of thorny branches, cloſely entangled, thereby ſecuring her retreat from the rude attacks of other birds; but it is not ſafety alone ſhe conſults, the inſide is furniſhed with a ſort of mattraſs compoſed of wool and other ſoft materials, on which her young repoſe: She lays ſeven or eight eggs, of a pale green colour, ſpotted with black.

The Magpie is crafty and familiar, and may be taught to pronounce words and even ſhort ſentences, and will imitate any particular noiſe which it hears. It is addicted, like other birds of its kind, to ſtealing, and will hoard up its proviſions. It is ſmaller than the Jackdaw, and its wings are ſhorter in proportion; accordingly its flight is not ſo lofty, nor ſo well ſupported: It never undertakes diſtant journies, but flies only from tree to tree, at moderate diſtances.

THE RED-LEGGED CROW.

CORNISH CHOUGH.

(*Corvus Graculus*, Lin.—*Le Coracias*, Buff.)

This bird is about the ſize of the Jack-daw: The bill is long, much curved, ſharp at the tip, and of a bright red colour; the iris of the eye is compoſed of two circles, the outer one red, the inner light blue; the eye-lids are red; the plumage is altogether of a purpliſh violet black; the legs are as red as the bill; the claws are large, much hooked, and black.

Buffon deſcribes this bird "as of an elegant figure, lively, reſtleſs and turbulent, but it may be tamed to a certain degree." It builds on high cliffs by the ſea ſide, and chiefly frequents the coaſts of Devonſhire and Cornwall, and likewiſe many parts of Wales; a few are found on the Dover cliffs, and ſome in Scotland. The female lays four or five white eggs, ſpotted with yellow. It is a voracious, bold, and greedy bird, and feeds on inſects and berries: It is ſaid to be particularly fond of the juniper berry. Its manners are like thoſe of a Jackdaw: It is attracted by glittering objects. Buffon ſays that it has been known to pull from the fire lighted pieces of wood, to the no ſmall danger of the houſe.

THE NUTCRACKER.

(*Corvus Caryocataƈtes*, Lin.—*Le Caſſe Noix*, Buff.)

The length of this bird is thirteen inches: The bill is about two inches long, and black; the eyes hazel; the upper part of the head and back part of the neck are black; its general colour is that of a duſky brown, covered with triangular ſpots of white; the wings are black; greater wing coverts tipped with white; the tail is white at the tip; the reſt black; rump white; legs and claws black.

There are very few inſtances known of this bird having been ſeen in England: It is common in Germany, is found alſo in Sweden and Denmark, and frequents the moſt mountainous parts of thoſe countries. It makes its neſt in holes of trees, and feeds on nuts, acorns, and the kernels of the pine apple. It is ſaid to pierce the bark of trees with its bill, like the Woodpecker. Our drawing was made from a ſtuffed ſpecimen in the muſeum of George Allan, Eſq.

THE JAY.

(*Corvus Glandarius*, Lin.—*Le Geai*, Buff.)

THIS moſt beautiful bird is not more than thirteen inches in length: Its bill is black; eyes white; the feathers on the forehead are white, ſtreaked with black, and form a tuft on its forehead, which it can erect at pleaſure; the chin is white, and from the corners of the bill on each ſide proceeds a broad ſtreak of black, which paſſes under the eye; the hinder part of the head, neck, and back are of a light cinnamon colour; the breaſt is of the ſame colour, but lighter; leſſer wing coverts bay; the belly and vent almoſt white; the greater wing coverts are elegantly barred with black, fine pale blue and white alternately; the greater quills are black, with pale edges, the baſes of ſome of them white; leſſer quills black; thoſe next the body cheſtnut; the rump is white; tail black, with pale brown edges; legs dirty pale brown.

The Jay is a very common bird in Great Britain, and is found in various parts of Europe. It is diſtinguiſhed as well for the beautiful arrangement of its colours, as for its harſh, grating voice, and reſtleſs diſpoſition. Upon ſeeing the ſportſman, it gives, by its cries, the alarm of danger, and thereby defeats his aim and diſappoints him.—The Jay builds in woods, and makes an artleſs neſt, compoſed of ſticks, fibres, and tender twigs: The female lays five or ſix eggs, of a greyiſh aſh colour, mixed with green, and faintly ſpotted with brown. Mr Pennant obſerves, that the young ones continue with their parents till the following ſpring, when they ſeparate to form new pairs. Birds of this ſpecies live on acorns, nuts, ſeeds, and various kinds of fruits; they will eat eggs, and ſometimes deſtroy young birds in the ab-

G

ſence of the old ones. When kept in a domeſtic ſtate they may be rendered very familiar, and will imitate a variety of words and ſounds. We have heard one imitate ſo exactly the ſound made by the action of a ſaw, that, tho' it was on a Sunday, we could hardly be perſuaded but that the perſon who kept it had a carpenter at work in the houſe.—A Jay, kept by a perſon we were acquainted with, at the approach of cattle, had learned to hound a cur dog upon them, by whiſtling and calling upon him by his name; at laſt, during a ſevere froſt, the dog was, by that means, excited to attack a cow big with calf, when the poor animal fell on the ice and was much hurt. The Jay being complained of as a nuiſance, its owner was obliged to deſtroy it.

THE CHATTERER.

SILK TAIL, OR WAXEN CHATTERER.

(*Ampelis Garrulus*, Lin.—*Le Jaſeur de Boheme*, Buff.)

THIS beautiful bird is about eight inches in length: Its bill is black, and has a ſmall notch at the end; its eyes, which are black and ſhining, are placed in a band of black, which paſſes from the baſe of the bill to the hind part of the head; its throat is black; the feathers on the head are long, forming a creſt; all the upper parts of the body are of a reddiſh aſh colour, the breaſt and belly inclining to purple; vent and upper tail coverts nearly white; the tail feathers are black, tipped with pale yellow; the quills are black, the third and fourth tipped on their outer edges with white, the five following with ſtraw colour; the ſecondaries with white, each being tipt or pointed with a flat horny ſubſtance of a bright vermillion colour. Theſe appendages vary in different ſubjects—in one of thoſe we had in our poſſeſſion, we counted eight on one wing and ſix on the other; the legs are ſhort and black. It is ſaid the female is not diſtinguiſhed by the little red waxen appendages at the ends of the ſecond quills; but this we are not able to determine from our own obſervations.

This rare bird viſits us only at uncertain intervals. In the year 1790 and 1791 ſeveral of them were taken in Northumberland and Durham as early as the month of November; ſince that time we have not heard of any being ſeen here. The ſummer reſidence of theſe birds is ſuppoſed to be the northern parts of Europe, within the arctic circle, from whence they ſpread themſelves into other countries, where they remain during winter, and return in the ſpring to their uſual haunts. The general food of this bird is berries of various kinds; in ſome countries it is ſaid to be extremely fond of grapes; one, which we ſaw in a ſtate of captivity was fed chiefly with quicken-tree berries, but from the difficulty of providing it with a ſufficient ſupply of its natural food it ſoon died. This is the only bird of its kind found in Europe; all the reſt are natives of America.

THE ROLLER.

(*Coracias Garrula*, Lin.—*Le Rollier d'Europe*, Buff.)

This rare bird is diſtinguiſhed by a plumage of moſt exquiſite beauty; it vies with the Parrot in an aſſemblage of the fineſt ſhades of blue and green, mixed with white, and heightened by the contraſt of graver colours, from whence perhaps it has been called the German Parrot, although in every other reſpect it differs from that bird, and rather ſeems to claim affinity with the Crow kind, to which we have made it an appendage. In ſize it refembles the Jay, being fomewhat more than twelve inches in length: Its bill is black, befet with fhort briftles at the bafe; the eyes are furrounded with a ring of naked fkin, of a yellow colour, and behind them there is a kind of wart; the head, neck, breaft, and belly are of a light pea green; the back and fcapulars reddifh brown; the points of the wings and upper coverts are of a rich deep blue, the greater coverts pale green; the quills are of a dufky hue, inclining to black, and mixed with deep blue; the rump is blue; the tail is fomewhat forked, the lower part of the feathers are of a dufky green, middle part pale blue, tips black; the legs are fhort, and of a dull yellow.—This is the only one of its kind found in Europe; it is very common in fome parts of Germany, but is fo rare in this country as hardly to deferve the name of a Britifh bird. The author of the Britifh Zoology mentions two that were fhot in England, and thefe we may fuppofe have been only ftragglers. Our drawing was made from a ftuffed fpecimen in the Mufeum of the late Mr Tunftall, at Wycliffe.

The Roller is wilder than the Jay, and frequents the thickeft woods; it builds its neft chiefly on birch trees. Buffon fays it is a bird of paffage, and migrates in the months of May and September. In thofe countries where it is common, it is faid to fly in large flocks in the autumn, and is frequently feen in cultivated grounds, with

Rooks and other birds, ſearching for worms, ſmall ſeeds, roots, &c.; it likewiſe ſeeds on berries, caterpillars, and inſects, and is ſaid, in caſes of neceſſity, to eat young frogs and even carrion. The female is deſcribed by Aldrovandus as differing very much from the male; its bill is thicker, and its head, neck, breaſt, and belly are of a cheſtnut colour, bordering on a greyiſh aſh. The young ones do not attain their brilliant colours till the ſecond year.

This bird is remarkable for making a chattering kind of noiſe, from whence it has obtained the name of Garrulus.

THE STARLING.

STARE.

(*Sturnus Vulgaris*, Lin.—*L'Etourneau*, Buff.)

The length of this bird is ſomewhat leſs than nine inches: The bill is ſtrait, ſharp-pointed, and of a yellowiſh brown—in old birds deep yellow; the noſtrils are ſurrounded by a prominent rim; the eyes are brown; the whole plumage is dark, gloſſed with blue, purple, and copper, each feather being marked at the end with a pale yellow ſpot, which is ſmaller and more numerous on the head and neck: the wing coverts are edged with yellowiſh brown; the quill and tail feathers duſky, with light edges; the legs are of a reddiſh brown.

From the ſtriking ſimilarity, both in form and manners, obſervable in this bird and thoſe more immediately preceding, we have no ſcruple in removing it from its uſual place, as it evidently forms a connecting link between them, and in a variety of points ſeems equally allied to both.—Few birds are more generally known than the Stare, being an inhabitant of almoſt every climate; and as it is a familiar bird, and eaſily trained in a ſtate of captivity, its habits have been more frequently obſerved than thoſe of moſt other birds. The female makes an artleſs neſt, in the hollows of trees, rocks, or old walls, and ſometimes in cliffs overhanging the ſea; ſhe lays four or five eggs, of a pale greeniſh aſh colour; the young birds are of a duſky brown colour till the firſt moult. In the winter ſeaſon theſe birds fly in vaſt flocks, and may be known at a great diſtance by their whirling mode of flight, which Buffon compares to a ſort of vortex, in which the collective body performs an uniform circular revolution, and at the ſame time continues to make a progreſſive advance. The evening is the time when the Stares aſſemble in the greateſt numbers, and betake themſelves to the fens and marſhes, where they rooſt among the reeds: They chatter much in the evening and morning, both when they aſſemble and diſperſe. So attached are they to ſociety, that they not only join thoſe of their own ſpecies, but alſo birds of a different kind, and are frequently ſeen in company with Redwings, Fieldfares, and even with Crows, Jackdaws, and Pigeons. Their principal food conſiſts of worms, ſnails, and caterpillars; they likewiſe eat various kinds of grain, ſeeds, and berries, and are ſaid to be particularly fond of cherries. In a confined ſtate they eat ſmall pieces of raw fleſh, bread ſoaked in water, &c. The Starling is very docile, and may eaſily be taught to repeat ſhort phraſes, or whiſtle tunes with great exactneſs, and in this ſtate acquires a warbling ſuperior to its native ſong.

THE ROSE-COLOURED OUZEL.

(*Turdus, Roſeus*, Lin.—*Le Merle Couleur de Roſe*, Buff.)

Is the ſize of a Starling: Its bill is of a carnation colour, blackiſh at the baſe; irides pale; the feathers on the head are long, forming a creſt; the head, neck, wings, and tail are black, gloſſed with ſhades of blue, purple, and green; its back, rump, breaſt, belly, and leſſer wing coverts pale roſe colour, marked with a few irregular dark ſpots; legs pale red; claws brown.

This bird has been ſo rarely met with in England that it will ſcarcely be admitted amongſt ſuch as are purely Britiſh. There are however a few inſtances of its being found here; and, although not a reſident, it ſometimes viſits us, on which account it muſt not be paſſed over unnoticed. It is found in various parts of Europe and Aſia, and in moſt places is migratory. It ſeems to delight moſt in the warmer climates; it is fond of locuſts, and frequents the places where thoſe deſtructive inſects abound in great numbers; on which account it is ſaid to be held ſacred by the inhabitants.

THE RING OUZEL.

(*Turdus Torquatus*, Lin.—*Le Merle à Plaſtron Blanc*, Buff.)

This bird very much reſembles the Blackbird: Its general colour is of a dull black or duſky hue, each feather being margined with a greyiſh aſh colour; the bill is duſky, corners of the mouth and inſide yellow; eyes hazel; its breaſt is diſtinguiſhed by a creſcent of pure white, which almoſt ſurrounds the neck, and from whence it derives its name; its legs are of a duſky brown. The female differs in having the creſcent on the breaſt much leſs conſpicuous, and in ſome birds wholly wanting, which has occaſioned ſome authors to conſider it as a different ſpecies, under the name of the Rock Ouzel.

Ring Ouzels are found in various parts of this kingdom, chiefly in the wilder and more mountainous parts of the country; its habits are ſimilar to thoſe of the Blackbird; the female builds her neſt in the ſame manner, and in ſimilar ſituations, and lays four or five eggs of the ſame colour: They feed on inſects and berries of various kinds, are fond of grapes, and, Buffon obſerves, during the ſeaſon of vintage are generally fat, and at that time are eſteemed delicious eating. The ſame author ſays, that in France they are migratory, and in ſome parts of this kingdom they have been obſerved to change places, particularly in Hampſhire, where they are known generally to ſtay not more than a fortnight at one time. Our repreſentation was taken from one killed near Bedlington in Northumberland.

THE BLACK OUZEL.

BLACKBIRD.

(*Turdus Merula*, Lin.—*Le Merle*, Buff.)

THE length of the Blackbird is generally about ten inches: Its plumage is altogether black; the bill, inſide of the mouth, and edges of the eye-lids are yellow, as are alſo the ſoles of the feet; the legs are of a dirty yellow. The female is moſtly brown, inclining to ruſt colour on the breaſt and belly; the bill is duſky, and the legs brown; its ſong is alſo very different, ſo that it has ſometimes been miſtaken for a bird of a different ſpecies.

Male Blackbirds, during the firſt year, reſemble the females ſo much as not eaſily to be diſtinguiſh-ed from them; but after that, they aſſume the yellow bill, and other diſtinguiſhing marks of their kind. The Blackbird is a ſolitary bird, frequenting woods and thickets, chiefly of evergreens, ſuch as pines, firs, &c. eſpecially where there are perennial ſprings, which afford it both ſhelter and ſubſiſtence. Wild Blackbirds feed on berries, fruits, inſects, and worms; they never fly in flocks like Thruſhes; they pair early, and begin to warble ſooner than any other of the ſongſters of the grove. The female builds her neſt in buſhes or low trees, and lays four or five eggs, of a blueiſh green colour, marked irregularly with duſky ſpots. The young birds are eaſily brought up tame, and may be taught to whiſtle a variety of tunes, for which their clear, loud, and ſpirited tones are well adapted. They are reſtleſs and timorous birds, eaſily alarmed, and difficult of acceſs; but Buffon obſerves that they are more reſtleſs than cunning, and more timorous than ſuſpicious, as they readily ſuffer themſelves to be caught with bird-lime, nooſes, and all ſorts of ſnares. They are never kept in aviaries; for when ſhut up with other birds they purſue and haraſs their companions in ſlavery unceaſingly, for which reaſon they are generally confined in cages apart. In ſome counties of England this bird is called the Ouzel.

MISSEL THRUSH.

MISSEL BIRD OR SHRITE.

(*Turdus Viſcivorus*, Lin.—*La Drainè*, Buff.)

The length of this bird is about eleven inches: The bill is duſky, the baſe of the lower bill yellow; the eyes hazel; the head, back, and leſſer coverts of the wings are of a deep olive brown, the latter tipped with white; the lower part of the back and rump tinged with yellow; the cheeks are of a yellowiſh white, ſpotted with brown, as are alſo the breaſt and belly, which are marked with larger ſpots, of a dark brown colour; the quills are brown, with pale edges; tail feathers the ſame: the three outermoſt tipped with white; the legs are yellow; claws black. The female builds her neſt in buſhes or low trees, and lays four or five eggs, of a dirty fleſh colour, marked with blood red ſpots. Its neſt is made of moſs, leaves, &c. lined with dry graſs, ſtrengthened on the outſide with ſmall twigs. It begins to ſing very early, often on the turn of the year in blowing ſhowery weather, from whence in ſome places it is called the Storm-cock. Its note of anger is very loud and harſh, between a chatter and a ſhriek, which accounts for ſome of its names. It feeds on various kinds of berries, particularly thoſe of the miſletoe, of which birdlime is made. It was formerly believed that the

the plant of that name was only propagated by the ſeed which paſſed the digeſtive organs of this bird, from whence aroſe the proverb "*Turdus malum ſibi cacat*;" it likewiſe feeds on caterpillars and various kinds of inſects, with which it alſo feeds its young. This bird is found in various parts of Europe, and is ſaid to be migratory in ſome places, but continues in England the whole year, and frequently has two broods.

THE FIELDFARE.

(*Turdus Pilaris*, Lin.—*La Litorne, ou Tourdelle*, Buff)

This is ſomewhat leſs than the Miſſel Thruſh; its length ten inches: The bill is yellow; each corner of the mouth is furniſhed with a few black briſtly hairs; the eye is light brown; the top of the head and back part of the neck are of a light aſh colour, the former ſpotted with black; the back and coverts of the wings are of a deep hoary brown; the rump aſh-coloured; the throat and breaſt are yellow, regularly ſpotted with black; the belly and thighs of a yellowiſh white; the tail brown, inclining to black; legs duſky yellowiſh brown; in young birds yellow.

The Fieldfare is only a viſitant in this iſland, making its appearance about the beginning of October, in order to avoid the rigorous winters of the North, from whence it ſometimes comes in great flocks, according to the ſeverity of the ſeaſon, and leaves us about the latter end of February or the beginning of March, and retires to Ruſſia, Sweden, Norway, and as far as Siberia and Kamſchatka. Buffon obſerves that they do not arrive in France till the beginning of December, that they aſſemble in flocks of two or three thouſand, and feed on ripe cervices, of which they are extremely fond: During the winter they feed on haws and other berries, they likewiſe eat worms, ſnails, and ſlugs.—Fieldfares ſeem of a more ſocial diſpoſition than the Throſtles or the Miſſels; they are ſometimes ſeen ſingly, but in general form very numerous flocks, and fly in a body, and though they often ſpread themſelves through the meadows in ſearch of food, they ſeldom loſe ſight of each other, but when alarmed fly off, and collect together upon the ſame tree. We have ſeen a variety of this bird, of which the head and neck were of a yellowiſh white; the reſt of the body was nearly of the ſame colour, mixed with a few brown feathers; the ſpots on the breaſt were faint and indiſtinct; the quill feathers were perfectly white, except one or two on each ſide, which were brown; the tail was marked in a ſimilar manner.

THE THROSTLE.

THRUSH OR MAVIS.

Turdus Musicus, Lin.—*La Grive*, Buff.)

This is larger than the Redwing, but much less than the Missel, to which it bears a strong resemblance both in form and colours; a small notch is observable at the end of the bill, which belongs to this and every bird of the Thrush kind; the throat is white, and the spots on the breast more regularly formed than those of the Missel Thrush, being of a conical shape; the inside of the wings and the mouth are yellow, as are also the legs; the claws are strong and black.—The Throstle is distinguished among our singing birds by the clearness and fullness of its note; it charms us not only with the sweetness, but variety of its song, which begins early in the ſpring, and continues during part of the ſummer. This bold and pleaſing ſongſter, from his high ſtation, ſeems to command the concert of the grove, whilſt, in the beautiful language of the poet,

"The Jay, the Rook, the Daw,
"And each harſh pipe (diſcordant heard alone)
"Aid the full concert, while the Stock-Dove breathes
"A melancholy murmur through the whole."

The female builds her neſt generally in buſhes; it is compoſed of dried graſs, with a little earth or clay intermixed, and lined with rotten wood; ſhe lays five or ſix eggs of a pale blue colour, marked with duſky ſpots. Although this ſpecies is not conſidered as migratory with us, it has, nevertheleſs, been obſerved in ſome places in great numbers during the ſpring and ſummer, where not one was to be ſeen in the winter, which has induced an opinion that they either ſhift their quarters entirely, or take ſhelter in the more retired parts of the woods.—That the Throſtle is migratory in France, we have the authority of that nice obſerver of nature, M. de Buffon, who ſays that it appears in Burgundy about the end of September, before the Redwing and Fieldfare, and that it feeds upon the ripe grapes, and ſometimes does much damage to the vineyard. The females of all the Thruſh kind are very ſimilar to the males, and differ chiefly in a leſſer degree of brilliancy in the colours.

THE REDWING.

SWINEPIPE OR WIND THRUSH.

Turdus Iliacus, Lin.—*Le Mauvis*, Buff.)

Is not more than eight inches in length: The bill is of a dark brown colour; eyes deep hazel; the plumage in general is ſimilar to that of the Thruſh, but a white ſtreak over the eye diſtinguiſhes it from that bird; the belly is not quite ſo much ſpotted, and the ſides of the body and under the wings are tinged with red, which is its peculiar characteriſtic, from whence alſo it derives its name.

Theſe birds make their appearance a few days before the Fieldfare,* and are generally ſeen with

* A Redwing was taken up November 7th, 1785, at ſix

them after their arrival; they frequent the ſame places, eat the ſame food, and are very ſimilar to them in manners. Like the Fieldfare it leaves us in the ſpring, for which reaſon its ſong is quite unknown to us, but it is ſaid to be very pleaſing. The female builds its neſt in low buſhes or hedges, and lays ſix eggs, of a greeniſh blue colour, ſpotted with black. This and the former are delicate eating; the Romans held them in ſuch eſtimation that they kept thouſands of them together in aviaries, and fed them with a ſort of paſte made of bruiſed figs and flour, and various other kinds of food to improve the delicacy and flavour of their fleſh: Theſe aviaries were ſo contrived as to admit light barely ſufficient to direct them to their food; every object which might tend to remind them of their former liberty was carefully kept out of ſight, ſuch as the fields, the woods, the birds, or whatever might diſturb the repoſe neceſſary for their improvement. Under this management theſe birds fattened to the great profit of their proprietors, who ſold them to Roman epicures for three denarii, or about two ſhillings ſterling each.

o'clock in the morning, which, on its approach to land, had flown againſt the light-houſe at Tynemouth, and was ſo ſtunned that it fell to the ground and died ſoon after; the light moſt probably had attracted its attention.

THE CUCKOO.

THE GOWK.

(*Cuculus Canorus*, Lin.—*Le Coucou*, Buff.)

Length fourteen inches; breadth twenty-five: Its bill is black, and ſomewhat bent; eyes yellow; inſide of the mouth red; its head, neck, back, and wing coverts are of a pale blue or dove colour, which is darkeſt on the head and back, and paleſt on the fore part of the neck and rump; its breaſt and belly are white, elegantly croſſed with wavy bars of black; the quill feathers are duſky, their inner webs marked with large oval white ſpots; the tail is long; the two middle feathers are black, with white tips; the others duſky, marked with alternate ſpots of white on each ſide the ſhaft; the legs are ſhort and of a yellow colour; toes two forward, two backward; claws white.

The Cuckoo viſits us early in the ſpring—its well-known cry is generally heard about the middle of April, and ceaſes the latter end of June; its ſtay is ſhort, the old Cuckoos being ſaid to quit this country early in July. Cuckoos never pair; they build no neſt; and, what is more extraordinary, the female depoſits her ſolitary egg in that of another bird, by whom it is hatched. The neſt ſhe chuſes for this purpoſe is generally ſelected from the following, viz. The Hedge-ſparrow, the Water-wagtail, the Titlark, the Yellow-hammer, the Green Linnet, or the Whinchat. Of theſe it has been obſerved that ſhe ſhews a much greater partiality to the Hedge-ſparrow than to any of the reſt.

We owe the following account of the economy of this ſingular bird in the diſpoſal of its egg, to the accurate obſervations of Mr Edward Jenner, communicated to the Royal Society, and publiſhed in the 78th volume of their tranſactions, part II. He obſerves that, during the time the Hedge-ſparrow is laying her eggs, which generally takes up four or five days, the Cuckoo contrives to depoſit her egg among the reſt, leaving the future care of it entirely to the Hedge-ſparrow. This intruſion often occaſions ſome diſcompoſure, for the old Hedge-ſparrow at intervals, whilſt ſhe is ſitting, not only throws out ſome of her own eggs, but ſometimes injures them in ſuch a way that they become addle, ſo that it frequently happens that not more than two or three of the parent bird's eggs are hatched with that of the Cuckoo; and what is very remarkable, it has never been obſerved that the Hedge-ſparrow has either thrown out or injured the egg of the Cuckoo. When the Hedge-ſparrow has ſat her uſual time, and diſengaged the young Cuckoo and ſome of her own offſpring from the ſhell, her own young ones, and any of her eggs that remain unhatched, are ſoon turned out, the young Cuckoo remaining in full poſſeſſion of the neſt, and the ſole object of the future care of her foſter parent. The young birds are not previouſly killed, nor the eggs demoliſhed, but all are left to periſh together, either entangled in the buſh which contains the neſt, or lying on the ground under it. Mr Jenner next proceeds to account for this ſeemingly unnatural circumſtance; and as what he has advanced is the reſult of his own repeated obſervations, we ſhall give it nearly in his own words. "On the 18th June, 1787, Mr J. examined the neſt of a Hedge-ſparrow, which then contained a Cuckoo's and three Hedge-ſparrow's eggs. On inſpecting it the day following, the bird had hatched, but the neſt then contained only a young Cuckoo and one young Hedge-ſparrow. The neſt was placed ſo near the extremity of a hedge that he could diſtinctly ſee what was going forward in it; and, to his great aſtoniſhment, he ſaw the young Cuckoo, though ſo lately hatched, in the act of turning out the young Hedge-ſparrow. The mode of accompliſhing this was curious: The little animal, with the aſſiſtance of its rump and wings, contrived to get the bird upon its back, and making a lodgement for its burden by elevating its elbows, clambered backwards with it up the ſide of the neſt till it reached the top, where reſting for a moment, it threw off its load with a jerk, and quite diſengaged it from the neſt: After remaining a ſhort time in this ſituation, and feeling about with the extremities of its wings, as if to be convinced that the buſineſs was properly executed, it dropped into the neſt again. Mr J. made ſeveral experiments in different neſts by repeatedly putting in an egg to the young Cuckoo, which he always found to be diſpoſed of in the ſame manner. It is very remarkable, that nature ſeems to have provided for the ſingular diſpoſition of the Cuckoo in its formation at this period, for, different from other newly hatched birds, its back from the ſcapulæ downwards is very broad, with a conſiderable depreſſion in the middle, which ſeems intended by nature for the purpoſe of giving a more ſecure lodgement to the egg of the Hedge-ſparrow, or its young one, while the young Cuckoo is employed in removing either of them from the neſt. When it is above twelve days old this cavity is quite filled up, the back aſſumes the ſhape of neſtling birds in general, and at that time the diſpoſition for turning out its companion entirely ceafes. The fmallnefs of the Cuckoo's egg, which, in general, is lefs than that of the Houfe fparrow,* is another circumftance to be attended to in this furprizing tranfaction, and feems to account for the parent Cuckoo's depofiting it in the nefts of fuch fmall birds only as have been mentioned. If fhe were to do this in the neft of a bird which produced a larger egg, and confequently a larger neftling, its defign would probably be fruftrated; the young Cuckoo would be unequal to the tafk of becoming fole poffeffor of the neft, and might fall a facrifice to the fuperior ftrength of its partners.

Mr Jenner obferves, that it fometimes happens that two Cuckoos' eggs are depofited in the fame neft, and gives the following inftance of one which fell under his obfervation. Two Cuckoos and a Hedge-fparrow were hatched in the fame neft, one Hedge-fparrow's egg remaining unhatched: In a few hours a conteft began between the Cuckoos for poffeffion of the neft, which continued undetermined till the afternoon of the following day, when one of them, which was fomewhat fuperior in fize, turned out the other, together with the young Hedge-fparrow and the unhatched egg. This conteft, he adds, was very remarkable: The combatants alternately appeared to have the advantage, as

* The Cuckoo eggs which have come under our obfervation were nearly of the fize of thofe of the Thrufh.

each carried the other ſeveral times nearly to the top of the neſt, and then ſunk down again oppreſſed with the weight of its burthen: till at length, after various efforts, the ſtrongeſt prevailed, and was afterwards brought up by the Hedge-ſparrow. It would carry us beyond the limits of our work to give a detail of the obſervations made by our ingenious inquirer; we muſt therefore refer our reader to the work itſelf, in which he will find a variety of matter entirely new reſpecting this ſingular bird, whoſe hiſtory has for ages been enveloped in fable, and mixed with unaccountable ſtories founded in ignorance and ſuperſtition. At what period the young Cuckoos leave this country is not preciſely known; Mr Jenner ſuppoſes they go off in ſucceſſion, and leave us as ſoon as they are capable of taking care of themſelves. That ſome of them remain here in a torpid ſtate we have already had occaſion to obſerve;* but this cannot be the caſe with the greater part of thoſe which leave this country and retire to milder climates, to avoid the rigours of winter. Buffon mentions ſeveral inſtances of young Cuckoos having been kept in cages, which, probably for want of proper nutriment, did not ſurvive the winter. We knew of one which was preſerved through the winter by being fed with worms, inſects, ſoaked bread, and ſmall pieces of fleſh. The plumage of the Cuckoo varies greatly at different

* See the introduction.

periods of its life. In young Cuckoos the bill, legs, and tail are nearly the ſame as in the old ones ; the eye is blue ; the throat, neck, breaſt, and belly are elegantly barred with a dark brown on a light ground ; the back is of a lead colour, mixed with brown, and faintly barred with white ; the tail feathers are irregularly marked with black, light brown, and white, and tipped with white ; its legs are yellow.

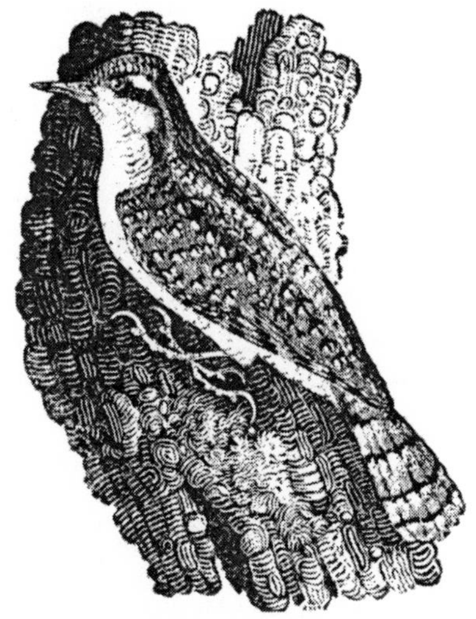

THE WRYNECK.

(*Jynx Torquilla*, Lin.—*Le Torcol*, Buff.)

The principal colours which diſtinguiſh this beautiful little bird conſiſt of different ſhades of brown, but ſo elegantly arranged as to form a picture of the moſt exquiſite neatneſs; from the back part of the head down to the middle of the back there runs an irregular line of dark brown, inclining to black; the reſt of the back is aſh-coloured, ſtreaked and powdered with brown; the throat and under ſide of the neck are of a reddiſh brown, croſſed with fine bars of black; the breaſt, belly, and thighs are of a light aſh colour, marked with triangular ſpots, irregularly diſperſed; the larger quill feathers are marked on the outer webs with alternate ſpots of dark brown and ruſt colour, which, when the wing is cloſed, give it the appearance of chequered work; the reſt of the wing and ſcapulars are nicely freckled and ſhaded with brown ſpots of different ſizes; the tail feathers are marked with irregular bars of black, the intervening ſpaces being finely freckled and powdered with dark brown ſpots; its bill is rather long, ſharp-pointed, and of a pale lead colour; its eyes are light brown; but what chiefly diſtinguiſhes this ſingular bird is the ſtructure of its tongue, which is of conſiderable length, of a cylindrical form, and capable of being puſhed forwards or drawn into its bill again; it is furniſhed with a horny ſubſtance at its end, with which it ſecures its prey and brings it to its mouth; its legs are ſhort and ſlender; the toes placed two before and two behind; the claws ſharp, much hooked, and formed for climbing the branches of trees, on which it can run in all directions with great facility. It makes an artleſs neſt, of dry graſs upon duſty rotten wood, in holes of trees, the entrance to which is ſo ſmall as ſcarcely to admit the hand, on which account its eggs are come at with difficulty; according to Buffon, they are perfectly white, and from eight to ten in number.—This curious bird, though ſimilar in many reſpects to the Woodpecker, ſeems to conſtitute a genus of itſelf: It is found in various parts of Europe, and generally appears with us a few days before the Cuckoo. Its food conſiſts chiefly of ants and other inſects, of which it finds great abundance lodged in the bark and crevices of trees. The ſtomach of one which we opened was full of indigeſted parts of ants. It is ſaid to frequent the places where ant hills are, into which it darts its tongue and draws out its prey. Though nearly related to the family of the Woodpeckers, in the formation of its bill and feet, it never aſſociates with them, but ſeems to form a ſmall and ſeparate family. The Wryneck holds itſelf very erect on the branch of the tree where it ſits; its body is almoſt bent backward, whilſt it writhes its head and neck by a ſlow and almoſt involuntary motion, not unlike the waving wreaths of a reptile. It is a very ſolitary bird, and leads a ſequeſtered life; it is never ſeen with any other ſociety but that of its female, and it is only tranſitory, for as ſoon as the domeſtic union is diſſolved, which is in the month of September, they retire and migrate by themſelves.

THE WOODPECKERS.

Of these only three or four kinds are found in these kingdoms. Their characters are striking and their manners singular. The bill is large, strong, and fitted for its employment; the end of it is formed like a wedge, with which it pierces the bark of trees and bores into the wood, in which its food is lodged. Its neck is short and thick, and furnished with powerful muscles, which enable it to strike with such force as to be heard at a considerable distance; its tongue is long and taper; at the end of it there is a hard bony substance, which penetrates into the crevices of trees, and extracts the insects and their eggs, which are lodged there; the tail consists of ten stiff, sharp-pointed feathers bent inwards, by which it secures itself on the trunks of trees while in search of food; for this purpose its feet are short and thick, and its toes, which are placed two forward and two backward, are armed with strong hooked claws, by which it clings firmly and creeps up and down in all directions. M. Buffon, with his usual warmth of imagination, thus describes the seemingly dull and solitary life of the Woodpecker.

" Of all the birds which earn their subsistence by " spoil, none leads a life so laborious and painful " as the Woodpecker: Nature has condemned it " to incessant toil and slavery. While others free-

" ly employ their courage or addrefs, and either " fhoot on rapid wing or lurk in clofe ambufh, the " Woodpecker is conftrained to drag out an infipid " exiftence in boring the bark and hard fibres of " trees to extract its humble prey. Neceffity ne- " ver fuffers any intermiffion of its labours, never " grants an interval of found repofe; often during " the night it fleeps in the fame painful pofture as " in the fatigues of the day. It never fhares the " fports of the other inhabitants of the air, it joins " not their vocal concerts, and its wild cries and " faddening tones, while they difturb the filence of " the foreft, exprefs conftraint and effort: Its " movements are quick, its geftures full of inquie- " tude, its looks coarfe and vulgar; it fhuns all fo- " ciety, even that of its own kind; and when it is " prompted to feek a companion, its appetite is not " foftened by delicacy of feeling."

THE GREEN WOODPECKER.

WOODSPITE, HIGH-HOE, HEW-HOLE, OR PICK-A-TREE.*

(*Picus Viridis*, Lin.—*Le Pic Verd*, Buff.)

This is the largeſt of the Britiſh kinds, being thirteen inches in length: Its bill is two inches long, of a triangular ſhape, and of a dark horn colour; the outer circle of the eye is white, ſurrounding another of red; the top of the head is of a bright crimſon, which extends down the hinder part of the neck, ending in a point behind; the eye

* Wallis, in his Hiſtory of Northumberland, obſerves that it is called by the common people Pick-a-tree, alſo Rain Fowl, from its being more loud and noiſy before rain. The old Romans called them *Pluviæ aves* for the ſame reaſon.

is ſurrounded by a black ſpace; and from each corner of the bill there is a crimſon ſtreak pointing downwards; the back and wing coverts are of an olive green; the rump yellow; the quill feathers are duſky, barred on the outer web with black and white; the baſtard wing is ſpotted with white; the ſides of the head and all the under parts of the body are white, ſlightly tinged with green; the tail is marked with bars like the wings; the legs are greeniſh. The female differs from the male in not having the red mark from the corner of the mouth; ſhe makes her neſt in the hollow of a tree, fifteen or twenty feet from the ground. Buffon obſerves that both male and female labour by turns in boring through the living part of the wood, ſometimes to a conſiderable depth, until they penetrate to that which is decayed and rotten, where ſhe lays five or ſix eggs, of a greeniſh colour, marked with ſmall black ſpots.

The Green Woodpecker is ſeen more frequently on the ground than the other kinds, particularly where there are ant-hills. It inſerts its long tongue into the holes through which the ants iſſue, and draws out theſe inſects in abundance. Sometimes, with its feet and bill, it makes a breach in the neſt, and devours them at its eaſe, together with their eggs. The young ones climb up and down the trees before they are able to fly; they rooſt very early, and repoſe in their holes till day.

GREATER SPOTTED WOODPECKER.

WITWALL.

(*Picus Major*, Lin.—*L'Epeiche, ou le Pic varie*, Buff.)

Its length is ſomewhat more than nine inches: The bill is of a dark horn colour, very ſtrong at the baſe; the upper and under ſides are formed by high-pointed ridges, which run along the middle of each; it is exceedingly ſharp at the end; the eyes are reddiſh, encircled with a large white ſpot, which extends to the back part of the head, on which there is a ſpot of crimſon; the forehead is buff colour; the top of the head black; on the back part of the neck there are two white ſpots, ſeparated by a line of black; the ſcapulars and tips of the wing coverts are white; the reſt of the plumage on the upper part of the body is black; the tail is black, the outer feathers marked with white ſpots; the throat, breaſt, and part of the belly are of a yellowiſh white; the vent and lower part of the belly crimſon; the legs and feet of a lead colour. The female wants the red ſpot on the back of the head.

This bird is common in England. Buffon ſays that it ſtrikes againſt the trees with briſker and harder blows than the Green Woodpecker:—It creeps with great eaſe in all directions upon the branches of trees, and is with difficulty ſeen, as it inſtantly avoids the ſight by creeping behind a branch, where it remains concealed.

THE MIDDLE-SPOTTED WOODPECKER.

(*Picus Medius*, Lin.—*Le Pic variè a tete Rouge*, Buff.)

This bird is ſomewhat leſs than the former, and differs from it chiefly in having the top of the head wholly crimſon; in every other reſpect the colours are much the ſame, though more obſcure. Buffon gives a figure of it in his *Planches Enluminees*, but conſiders it as a variety only of the former.

LESSER SPOTTED WOODPECKER.

HICKWALL.

(*Picus Minor*, Lin.—*Le petit Epeiche*, Buff.)

THIS is the ſmalleſt of our ſpecies, being only five inches and a half in length; weight nearly one ounce: Its general plumage is very ſimilar to the larger ſpecies, but wants the red under the tail, and the large white patches on the ſhoulders; the under parts of the body are of a dirty white; the legs lead colour. Buffon ſays, that in winter it draws near houſes and vineyards, that it neſtles like the former in holes of trees, and ſometimes diſputes poſſeſſion with the colemouſe, which it compels to give up its lodging.

THE NUTHATCH.

NUTJOBBER, WOODCRACKER.

(*Sitta Europea*, Lin.—*La Sittelle ou le Torchepot*, Buff.)

Its length is nearly ſix inches: The bill ſtrong, black above, beneath almoſt white; the eyes hazel; a black ſtroke paſſes over each eye, from the bill extending down the ſide of the neck as far as the ſhoulder; all the upper part of the body is of a fine blue grey colour; the cheeks and chin are white; breaſt and belly of a pale orange colour; ſides marked with ſtreaks of cheſtnut; quills duſky; its tail is ſhort, the two middle feathers are grey, the reſt duſky, three of the outermoſt ſpotted with white; the legs pale yellow; the claws large, ſharp, and much bent, the back claw very ſtrong; when extended, the foot meaſures one inch and three quarters.

This, like the Woodpecker, frequents woods, and is a ſhy and ſolitary bird; the female lays her eggs in holes of trees, frequently in thoſe which have been deſerted by the Woodpecker. During the time of incubation ſhe is aſſiduouſly attended by the male, who ſupplies her with food; ſhe is eaſily driven from her neſt, but on being diſturbed hiſſes like a ſnake. The Nuthatch feeds on caterpillars, beetles, and various kinds of inſects; it likewiſe eats nuts, and is very expert in cracking them ſo as to come at the contents; having placed a nut faſt in a chink, it takes its ſtand a little above, and ſtriking it with all its force, breaks the ſhell and catches up the kernel. Like the Woodpecker, it moves up and down the trunks of trees with great facility in ſearch of food. It does not migrate, but in the winter approaches nearer inhabited places, and is ſometimes ſeen in orchards and gardens. The young ones are eſteemed very good eating.

THE HOOPOE.

(*Upupa Epops*, Lin.—*Le Hupe ou Puput*, Buff.)

Its length is twelve inches, breadth nineteen: The bill is above two inches long, black, ſlender, and ſomewhat curved; the eyes hazel; the tongue very ſhort and triangular; the head is ornamented with a creſt, conſiſting of a double row of feathers of a pale orange colour, tipped with black, the higheſt about two inches in length; the neck is of a pale reddiſh brown; breaſt and belly white, which in young birds are marked with various duſky lines pointing downwards; the back, ſcapulars, and wings are croſſed with broad bars of black and white; the leſſer coverts of the wings light brown; the rump is white; the tail conſiſts of ten feathers, each marked with white, which, when clofed, affumes the form of a crefcent, the horns pointing downwards; the legs are fhort and black.

This is the only one of its kind found in thefe kingdoms; it is not very common with us, being feen only at uncertain periods. Our reprefentation was taken from a very fine one fhot near Bedlington, Northumberland, and fent us by the Rev. Mr Cotes. In its ftomach were found the claws and other indigeftible parts of infects of the beetle tribe; it was alive fome time after being fhot, and walked about erecting its tail and creft in a very pleafing manner. The female is faid to have two or three broods in the year; fhe makes no neft, but lays her eggs, generally about four or five in number, in the hollow of a tree, and fometimes in a hole in the wall, or even on the ground. Buffon fays, that he has fometimes found a foft lining of mofs, wool, or feathers in the nefts of thefe birds, and fuppofes that, in that cafe, they may have ufed the deferted neft of fome other bird. Its food confifts chiefly of infects, with the remains of which its neft is fometimes fo filled as to become extremely offenfive. It is a folitary bird, two of them being feldom feen together; in Egypt, where they are very common, they are feen only in fmall flocks. Its creft ufually falls behind on its neck, except when it is furprifed or irritated, and it then ftands erect.

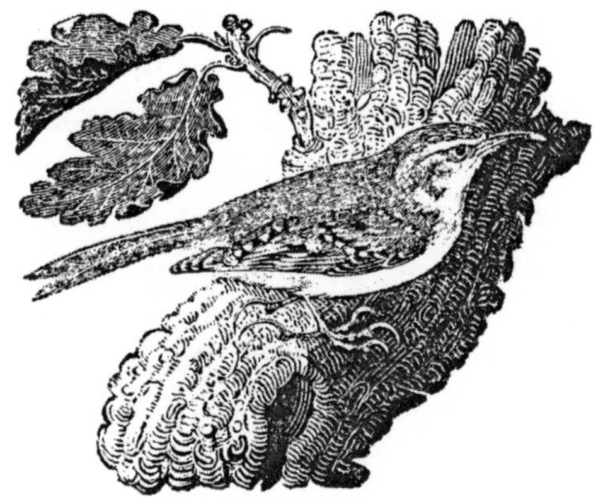

THE CREEPER.

(*Certhia familiaris*, Lin.—*Le Grimpereau*, Buff)

Its length is five inches and a half; the body is about the ſize of that of the Wren: Its bill is long, ſlender, and much curved, the upper one brown, the lower whitiſh; eyes hazel; the head, neck, back, and wing coverts are of a dark brown, variegated with ſtreaks of a lighter hue; the throat, breaſt, and belly are of a ſilvery white; the rump tawny; the quills are duſky, edged with tawny, and marked with bars of the ſame colour; the tips are white; above each eye a ſmall dark line paſſes towards the neck, above which there is a line of white; the tail is long, and conſiſts of twelve ſtiff feathers, of a tawny colour, pointed and forked at the end; the legs are ſhort and of a brown colour; the claws are long, ſharp, and much hooked, which enable it to run with great facility on all ſides of ſmall branches of trees in queſt of inſects and their eggs, which conſtitute its food. Although very common, it is not ſeen without difficulty, from the eaſe with which, on the appearance of any one, it eſcapes to the oppoſite ſide of the tree. It builds its neſt early in the ſpring, in the hole of a tree: The female lays from five to ſeven eggs, of an aſh colour, marked at the end with ſpots of a deeper hue.

OF THE PASSERINE ORDER.

This numerous claſs conſtitutes the fifth order in Mr Pennant's arrangement of Britiſh birds, and includes a great variety of different kinds: Of theſe we have detached the Stare, the Thruſh, and the Chatterer, and have joined them to the Pies, to which they ſeem to have a greater affinity. Thoſe which follow are diſtinguiſhed by their lively and active diſpoſitions, their beautiful plumage, and delightful melody. Of this order conſiſt thoſe amazing flocks of ſmall birds of almoſt every deſcription—thoſe numerous families, which, univerſally diffuſed throughout every part of the known world, people the woods, the fields, and even the largeſt and moſt populous cities, in countleſs multitudes, and every where enliven, diverſify, and adorn the face of nature. Theſe are not leſs conſpicuous for their uſefulneſs, than their numbers and variety: They are of infinite advantage in the economy of nature, in deſtroying myriads of noxious inſects, which would otherwiſe teem in every part of the animal and vegetable ſyſtems, and would pervade and choke up all the avenues of life and health. Inſects and their eggs, worms, berries, and ſeeds of almoſt every kind, form the varied maſs from whence theſe buſy little tribes derive their ſupport.

The characters of the Paſſerine order, which are as various as their habits and diſpoſitions, will be beſt ſeen in the deſcription of each particular kind. It may be neceſſary however to obſerve, that they naturally divide themſelves into two diſtinct kinds, namely, the hard-billed or ſeed birds, and the ſlender or ſoft-billed birds; the former are furniſhed with ſtout bills of a conical ſhape, and very ſharp at the point, admirably fitted for the purpoſe of breaking the hard external coverings of the ſeeds of plants from the kernels, which conſtitute the principal part of their food; the latter are remarkable for the ſoftneſs and delicacy of their bills; their food conſiſts altogether of ſmall worms, inſects, the larvæ of inſects and their eggs, which they find depoſited in immenſe profuſion on the leaves and bark of trees, in chinks and crevices of ſtones, and even in ſmall maſſes on the bare ground, ſo that there is hardly a portion of matter that does not contain a plentiful ſupply of food for this diligent race of beings.

" Full nature ſwarms with life;
" The flowery leaf
" Wants not its ſoft inhabitants. Secure
" Within its winding citadel, the ſtone
" Holds multitudes. But chief the foreſt-boughs,
" That dance unnumber'd to the playful breeze,
" The downy orchard, and the melting pulp
" Of mellow fruit, the nameleſs nations feed
" Of evaneſcent inſects."

OF THE GROSBEAK.

This genus is not numerous in theſe kingdoms, and of thoſe which we call ours, moſt of them are only viſitors, making a ſhort ſtay with us, and leaving us again to breed and rear their young in other countries. They are in general ſhy and ſolitary, living chiefly in woods at a diſtance from the habitations of men. Their vocal powers are not great; and as they do not add much to the general harmony of the woods which they inhabit, they are conſequently not much known or ſought after. Their moſt conſpicuous character conſiſts in the thickneſs and ſtrength of their bills, which enables them to break the ſtones of various kinds of fruits, and other hard ſubſtances on which they feed. Their general appearance is very ſimilar to birds of the Finch kind, of which they may be reckoned the principal branch.

THE CROSS-BILL.

SHEL-APPLE.

(*Loxia Curviroſtra*, Lin.—*Le bec Croiſé*, Buff.)

THIS bird is about the ſize of a Lark, being nearly ſeven inches in length: It is diſtinguiſhed by the peculiar formation of its bill, the upper and under mandible curving in oppoſite directions, and croſſing each other at the points;* its eyes are ha-

* This ſingular conſtruction of the bill is conſidered by M. Buffon as a defect or error in nature, rather than a permanent feature, merely becauſe that, in ſome ſubjects, the bill croſſes to the left, and in others to the right, ariſing, as he ſuppoſes, from the way in which the bird has been accuſtomed to uſe its bill, by employing either the one ſide or the other to lay hold of its

zel; its general colour is reddiſh, mixed with brown on the upper parts, the under parts are conſiderably paler, being almoſt white at the vent; the wings are ſhort, not reaching farther than the ſetting on of the tail—they are of a brown colour; the tail is of the ſame colour, and ſomewhat forked; the legs are black; the colours of the Croſsbill are extremely ſubject to variation; amongſt a great number there are hardly two of them exactly

food. This mode of reaſoning, however, muſt prove very defective, when we conſider that this peculiarity is confined to a ſingle ſpecies, no other bird in nature being ſubject to a ſimilar variation from the general conſtruction, although there are many other birds which feed upon the ſame kinds of hard ſubſtances, which, neverthelefs, do not experience any change in the formation and ſtructure of their bills; neither has the argument, drawn from the ſuppoſed exuberance of growth in the bills of theſe birds, any better foundation, as that likewiſe may be applied to other birds, and the ſame queſtion will occur—namely, Why is not the ſame effect produced? This ingenious but fanciful writer, in the further proſecution of his argument, ſeems to increaſe the difficulties in which it is involved. He obſerves, "that the bill, hooked upwards and downwards, and bent in oppoſite directions, ſeems to have been formed for the purpoſe of detaching the ſcales of the fir cones and obtaining the ſeeds lodged beneath them, which are the principal food of the bird. It raiſes each ſcale with its lower mandible, and breaks it with the upper." We think there needs no ſtronger argument than this to prove, that Nature, in all her operations, works by various means; and although theſe are not always clear to our limited underſtandings, the good of all her creatures is the one great end to which they are all directed.

ſimilar; they likewiſe vary with the ſeaſon and according to the age of the bird. Edwards paints the male with a roſe colour, and the female with a yellowiſh green, mixed more or leſs with brown. Both ſexes appear very different at different times of the year.

The Croſs-bill is an inhabitant of the colder climates, and has been found as far as Greenland. It breeds in Ruſſia, Sweden, Poland, and Germany, in the mountains of Switzerland, and among the Alps and Pyrenees, from whence it migrates in vaſt flocks into other countries. It ſometimes is met with in great numbers in this country, but its viſits are not regular,* as in ſome years it is rarely to be ſeen. Its principal food is ſaid to be the ſeeds of the pine tree; it is obſerved to hold the cone in one claw like the Parrot, and when kept in a cage has all the actions of that bird, climbing by means of its hooked bill, from the lower to the upper bars of its cage. From its mode of ſcrambling and the beauty of its colours, it has been called by ſome the German Parrot. The female is ſaid to begin to build as early as January; ſhe places her neſt under the bare branches of the pine tree, fixing it with the reſinous matter which exudes from that tree, and beſmearing it on the outſide with the ſame ſubſtance, ſo that the melted ſnow or rain cannot penetrate it.

* We have met with it on the top of Blackſton-edge, between Rochdale and Halifax, in the month of Auguſt.

THE GROSBEAK.

HAWFINCH.

(*Loxia Coccothrauſtes*, Lin.—*Le Gros-bec*, Buff.)

Length near ſeven inches: Bill of a horn colour, conical, and prodigiouſly thick at the baſe; eyes aſh-coloured; the ſpace between the bill and the eye, and from thence to the chin and throat, is black; the top of the head is of a reddiſh cheſtnut, as are alſo the cheeks, but ſomewhat paler; the back part of the neck is of a greyiſh aſh colour; the back and leſſer wing coverts cheſtnut; the greater wing coverts are grey, in ſome almoſt white, forming a band acroſs the wing; the quills are all black, except ſome of the ſecondaries neareſt the body, which are brown; the four outer quills ſeem as if cut off at the ends; the prime quills have each of them a ſpot of white about the middle of the inner web; the breaſt and belly are of a pale ruſt colour, growing almoſt white at the vent; the tail is black, except the ends of the middle feathers, which are grey; the outer ones are tipped with white; the legs are pale brown. The female greatly reſembles the male, but her colours are leſs vivid, and the ſpace between the bill and the eye is grey inſtead of black. Theſe birds vary conſiderably, ſcarcely two of them being alike: In ſome the head is wholly black, in others the whole upper part of the body is of that colour, and others have been met with entirely white, except the wings.

This ſpecies is an inhabitant of the temperate climates, from Spain, Italy, and France, as far as Sweden, but only viſits this iſland occaſionally, and generally in winter; probably being driven over in its paſſage from its northern haunts, to the milder climates of France and Italy. It breeds in theſe countries, but is no where numerous. Buffon ſays it is a ſhy and ſolitary bird, with little or no ſong; it generally inhabits the woods during ſummer, and in winter reſorts near the hamlets and farms. The female builds her neſt in trees; it is compoſed of ſmall dry roots and graſs, and lined with warmer materials: The eggs are roundiſh, of a blueiſh green, ſpotted with brown. She feeds her young with inſects, chryſalids, and other ſoft nutritious ſubſtances.

THE PINE GROSBEAK.

GREATEST BULLFINCH.

(*Loxia Enucleator*, Lin.—*Le Dur-bec*, Buff.)

This exceeds the laſt in ſize, being nine inches in length: The bill is duſky, very ſtout at the baſe, and ſomewhat hooked at the tip; the head, neck, breaſt, and rump are of a roſe coloured crimſon; the back and leſſer wing coverts black, each feather edged with reddiſh brown; the greater wing coverts tipped with white, forming two bars on the wing; the quills are black, with pale edges; the ſecondaries the ſame, but edged with white; the belly and vent are ſtraw-coloured; the tail is marked as the quills, and is ſomewhat forked: the legs are brown.

This bird is found only in the northern parts of this iſland and of Europe; it frequents the pine foreſts, and feeds on the ſeeds of that tree, like the Croſs-bill: It is likewiſe common in various parts of North America, viſiting the ſouthern ſettlements in the winter, and retiring northwards in the ſummer for the purpoſe of breeding. The female makes its neſt on trees, at a ſmall diſtance from the ground, and lays four white eggs, which are hatched in June.

GREEN GROSBEAK.

GREEN FINCH, OR GREEN LINNET.

(*Loxia Chloris*, Lin.—*Le Verdier*, Buff.)

THE bill is of a pale reddiſh brown or fleſh colour; eyes dark; the plumage in general is of a yellowiſh green; the top of the head, neck, back, and leſſer coverts olive green; the greater coverts and outer edges of the ſecondary quills aſh-coloured; the vent and tail coverts the ſame, daſhed with yellow; the rump yellow.

This bird is common in every part of Great Britain, and may be ſeen in almoſt every hedge, eſpecially during winter, when flocks of them keep together. The female makes her neſt in hedges or low buſhes; it is compoſed of dry graſs, and lined with hair, wool, and other warm materials; ſhe lays five or ſix eggs, of a pale greeniſh colour, marked at the larger end with ſpots of a reddiſh brown; ſhe is ſo cloſe a ſitter, that ſhe may ſometimes be taken on her neſt. The male is very attentive to his mate during the time of incubation, and takes his turn in ſitting. Though not diſtinguiſhed for its ſong, this bird is ſometimes kept in a cage, and ſoon becomes familiar. It does not migrate, but in the northern parts of our iſland it is ſeldom ſeen in winter, changing its quarters according to the ſeaſon of the year.

BULLFINCH.

ALP OR NOPE.

(*Loxia Pyrrhula*, Lin.—*Le Bouvreuil*, Buff.)

The bill is duſky; eyes black; the upper part of the head, the ring round the bill, and the origin of the neck, are of a fine gloſſy black;* the back aſh colour; the breaſt and belly red; wings and tail black; the upper tail coverts and vent are white; legs dark brown. The female is very ſimilar, but the colours in general are leſs bright, and the under parts of a reddiſh brown.†

* Hence in ſome countries it is called *Monk* or *Pope*, and in Scotland it is not improperly denominated *Coally hood*.

† The Bullfinch ſometimes changes its plumage, and becomes wholly black during its confinement, eſpecially when fed with hemp ſeed. In the Leverian Muſeum there is a variety of the Bullfinch entirely white.

This bird is common in every part of this iſland, as well as moſt parts of Europe; its uſual haunts, during ſummer, are in woods and thickets, but in winter it approaches nearer to cultivated grounds, and feeds on ſeeds, winter berries, &c.; in the ſpring it frequents gardens, where it is uſefully buſy in deſtroying the worms which are lodged in the tender buds. The female makes her neſt in buſhes; it is compoſed chiefly of moſs; ſhe lays five or ſix eggs, of a dull blueiſh white, marked at the larger end with dark ſpots. In a wild ſtate, its note is very ſimple; but when kept in a cage, its ſong, though low, is far from being unpleaſant. Both male and female may be taught to whiſtle a variety of tunes, and there are inſtances of two Bullfinches having been taught to ſing in parts; a wonderful inſtance of docility! They are frequently imported into this country from Germany, where they are taught to articulate, with great diſtinctneſs, ſeveral words.

OF THE BUNTING.

THE principal difference between this ſpecies and the laſt conſiſts in the formation of the bill, which in the Bunting is of a very ſingular conſtruction. The two mandibles are moveable, and the edges of each bend inwards; the opening of the mouth is not in a ſtreight line as in other birds, but at the baſe the junction is formed by an obtuſe angle in the lower mandible, nearly one third of its length, which is received by a correſponding angle in the upper one; in the laſt there is a hard knob, of great uſe in breaking the harder kinds of ſeeds and kernels, on which it feeds. The tongue is narrow, and tapers to a point like a tooth-pick; the firſt joint of the outer toe is joined to that of the middle one.

THE BUNTING.

(*Emberiza miliaria,* Lin.—*Le Proyer,* Buff.)

The length of this bird is about ſeven inches and a half: The bill is brown; iris hazel; the general colour reſembles that of a lark; the throat is white, the upper parts olive brown, each feather ſtreaked down the middle with black; the under parts are of a dirty yellowiſh white, ſtreaked on the ſides with dark brown, and ſpotted with the ſame on the breaſt; the quills are duſky, with yellowiſh edges; upper coverts tipped with white; tail feathers much the ſame as the wings, and ſomewhat forked; the legs pale brown.

This bird is very common in all parts of the country, and may be frequently obſerved on the higheſt part of the hedge or uppermoſt branch of a tree, uttering its harſh and diſſonant cry, which it inceſſantly repeats at ſhort intervals; this continues during the greateſt part of ſummer, after which they are ſeen in great flocks, and continue ſo for the moſt part during winter; they are often ſhot in great numbers, or caught in nets, and, from the ſimilarity of their plumage, are not unfrequently ſold for Larks. The female makes her neſt among the thick graſs, a little elevated above the ground: ſhe lays five or ſix eggs, and while ſhe is employed in the buſineſs of incubation, her mate brings her food, and entertains her with his frequently repeated ſong. Buffon obſerves, that in France the Bunting is ſeldom ſeen during winter, but that it migrates ſoon after the Swallow, and ſpreads itſelf through almoſt every part of Europe. Their food conſiſts chiefly of grain; they likewiſe eat variety of inſects, which they find in the fields and meadows.

YELLOW BUNTING.

YELLOW HAMMER, OR YELLOW YOWLEY.

(*Emberiza citrinella*, Lin.—*Le Bruant*, Buff.)

Length somewhat above six inches: Bill dusky; eyes hazel; its prevailing colour is yellow, mixed with browns of various shades; the crown of the head, in general, is bright yellow, more or less variegated with brown; the cheeks, throat, and lower part of the belly are of a pure yellow; the breast reddish, and the sides dashed with streaks of the same colour; the hind part of the neck and back are of a greenish olive; the greater quills are dusky, edged with pale yellow; lesser quills and scapulars dark brown, edged with grey; the tail is dusky, and a little forked, the feathers edged with light brown, the outermoſt with white; the legs are of a yellowiſh brown. It is ſomewhat difficult to deſcribe a ſpecies of bird of which no two are to be found perfectly ſimilar, but its ſpecific characters are plain, and cannot eaſily be miſtaken; the colours of the female are leſs bright than thoſe of the male, with very little yellow about the head.

This bird is common in every lane and on every hedge throughout the country, flitting before the traveller as he paſſes along the road, or uttering its ſimple and frequently repeated monotone on the hedges by the way ſide. They feed on various kinds of ſeeds, inſects, &c. The female makes an artleſs neſt, compoſed of hay, dried roots, and moſs, which ſhe lines with hair and wool; ſhe lays four or five eggs, marked with dark irregular ſtreaks, and frequently has more than one brood in the ſeaſon. In Italy, where ſmall birds of almoſt every deſcription are made uſe of for the table, they are eſteemed as very good eating, and are frequently fatted for that purpoſe like the Orlotan; but with us, who are accuſtomed to groſſer kinds of food, they are conſidered as too inſignificant to form any part of our repaſts.

THE BLACK-HEADED BUNTING.

REED BUNTING, OR REED SPARROW.

(*Emberiza Schæniclus*, Lin.—*L'Ortolan de Roſeaux*, Buff.)

This bird is about the ſize of the Yellow Bunting: Its eyes are hazel; the head, throat, fore part of the neck, and breaſt are black, which is divided by a white line from each corner of the bill, paſſing downward a little, and meeting on the back part of the neck, which it almoſt encircles; the upper parts of the body and wings are of a reddiſh brown, with a ſtreak of black down the middle of each feather; the under part of the body is white, with browniſh ſtreaks on the ſides; the rump and upper tail coverts blueiſh aſh colour, mixed with brown; the quills are duſky, edged with brown; the two middle feathers of the tail are black, with pale brown edges; the reſt wholly black, except the two outer ones, which are almoſt white, the ends tipped with brown, and the baſes black; the legs and feet duſky brown. The female has no collar; its throat is not ſo black, and its head is variegated with black and ruſt colour; the white on its under parts is not ſo pure, but is of a reddiſh caſt.

Birds of this ſpecies frequent fens and marſhy places, where there are abundance of ruſhes, among which it neſtles. The neſt is compoſed of dry graſs, and lined with the ſoft down of the reed; it is fixed with great art between four reed ſtalks, two on each ſide, almoſt cloſe to each other, and about three feet above the water: The female lays four or five eggs, of a pale blueiſh white, veined irregularly with purple, principally at the larger end. As its chief reſort is among the reeds, it is ſuppoſed that the ſeeds of that plant are its principal food; it is however frequently ſeen in the higher grounds near the roads, and ſometimes in corn fields. Theſe birds in general ſeek their food, ſimilar to the Bunting, in cultivated places; they keep near the ground, and ſeldom perch except among the buſhes. The male, during the time of hatching, has a ſoft, melodious, warbling ſong, whilſt it ſits perched among the reeds, and is frequently heard in the night time. It is a watchful, timorous bird, and is very eaſily alarmed; in a ſtate of captivity it ſings but little, and only when perfectly undiſturbed.

Birds of this ſpecies are migratory in France; with us they remain the whole year, and are ſeldom ſeen in flocks of more than three or four together. The one from whence our figure was taken was caught during a ſevere ſtorm in the midſt of winter.

SNOW BUNTING.

SNOWFLAKE.

(*Emberiza Nivalis*, Lin.—*L'Ortolan de Neige*, Buff.)

LENGTH near ſeven inches: Bill and eyes black; in winter the head, neck, coverts of the wings, rump, and all the under parts of the body are as white as ſnow, with a light tint of ruſty colour on the hind part of the head; the back is black; the baſtard wings and ends of the greater coverts white; the prime quills are black, ſecondaries white, with a black ſpot on their inner webs; middle feathers of the tail black, the three outer ones white, with a duſky ſpot near the ends; legs black. Its ſummer dreſs is different, the head, neck, and under parts of the body being marked with tranſverſe waves of a ruſty colour, of various ſtrength, but never ſo deep as in the female, of which it is the predominant colour; the white likewiſe upon the under parts of her body is leſs pure than that of the male.

The hoary mountains of Spitzbergen, the Lapland Alps, the ſhores of Hudſon's Bay, and perhaps countries ſtill more northerly, are, during the ſummer months, the favorite abodes of this hardy bird. The exceſſive ſeverity of theſe inhoſpitable regions changes parts of its plumage into white in winter; and there is reaſon to believe that the further northward they are found, the whiter the plumage will be. It is chiefly met with in the northern parts of this iſland, where it is called the Snowflake; it appears in great flocks in the ſnowy ſeaſon, and is ſaid to be the certain harbinger of ſevere weather, which drives it from its uſual haunts. This bird has been caught in various parts of Yorkſhire, and is frequently met with in Northumberland; it is found in all the northern latitudes without exception, as far as our navigators have been able to penetrate, great flocks of theſe birds having been ſeen by them upon the ice near the ſhores of Spitzbergen. They are known to breed in Greenland, where the female makes its neſt in the fiſſures of the mountain rocks; the outſide is compoſed of graſs, within which is a layer of feathers, and the down of the arctic fox compoſes the lining of its comfortable little manſion; ſhe lays five white eggs, ſpotted with brown. Theſe birds do not perch, but continue always on the ground, and run about like Larks, to which they are ſimilar in ſize, manners, and in the length of their hind claws, from whence they have been ranged with birds of that claſs by ſome authors, but with more propriety have been referred to the Buntings, from the peculiar ſtructure of their bill. They are ſaid to ſing ſweetly, ſitting on the ground. On their firſt arrival in this country they are very lean; but ſoon grow fat, and are conſidered as delicious food. The Highlands of Scotland abound with them.

TAWNY BUNTING.

GREAT PIED MOUNTAIN FINCH, OR BRAMBLING.

The length is ſomewhat above ſix inches: The bill is ſhort, of a yellow colour, and blackiſh at the point; the crown of the head tawny; the forehead cheſtnut colour; the hind part of the neck and cheeks the ſame, but paler; the throat, ſides of the neck, and ſpace round the eyes are of a dirty white; the breaſt dull yellow; the under parts white, in ſome tinged with yellow; the back and ſcapulars are black, edged with reddiſh brown; the quill feathers are duſky, edged with white; the ſecondaries are white on their outer edges; the greater coverts are tipped with white, which, when the wing is cloſed, forms a bed of white upon it; the upper tail coverts are yellow; the tail is a little forked, the two outermoſt feathers are white, the third black, tipped with white, the reſt wholly black; the legs are ſhort and black; the hind claws almoſt as long, but more bent than thoſe of the Lark.

Our figure and deſcription of this bird are taken from one which was caught in the high moory grounds above Shotley-Kirk, in the county of Northumberland. We are perfectly of opinion, with Mr Pennant, that this and the former are the ſame bird in their ſummer and winter dreſs.* Linnæus, who muſt have been well acquainted with this ſpecies, compriſes them under one, and ſays that they vary, not only from the ſeaſon, but according to their age: It is certain that no birds of the ſame ſpecies differ from each other more than they; amongſt multitudes, that are frequently taken, ſcarcely two being alike. Mr Pennant ſuppoſes, with great probability, that the ſwarms which annually viſit the northern parts of our iſland arrive from Lapland and Iceland, and make the iſles of Ferro, Shetland, and the Orkneys, their reſting-places during the paſſage. In the winter of 1778—9 they came in ſuch multitudes into Birſa, one of the Orkney iſles, as to cover the whole barony; yet, of all the numbers, it could hardly

* Vide Arctic Zoology, Number 222.

be diſcovered that any two of them agreed perfectly in colours. It is probable that the Mountain Bunting, or Leſſer Mountain Finch of Pennant and Latham, is the ſame bird in a ſomewhat different dreſs; it has been ſometimes found in the more ſouthern parts of England, where the little ſtranger would be noticed, and without duly attending to its diſtinguiſhing characters, has been conſidered as forming a diſtinct kind, and adding one more to the numerous varieties of the feathered tribes.—We have frequently had occaſion to obſerve, how difficult it is to avoid falling into errors of this ſort; the changes which frequently take place in the ſame bird, at different periods of its age, as well as from change of food, climate, or the like, are ſo conſiderable, as often to puzzle, and ſometimes to miſlead, the moſt experienced ornithologiſt; much caution is therefore neceſſary to guard againſt theſe deceitful appearances; leſt, by multiplying the ſpecies beyond the bounds which nature has preſcribed, we thereby introduce confuſion into our ſyſtem; and, inſtead of ſatisfying the attentive inquirer, we ſhall only bewilder and perplex him in his reſearches into nature.

OF THE FINCH.

The tranſition from the Bunting to the Finch is very eaſy, and the ſhade of difference between them, in ſome inſtances, almoſt imperceptible; on which account they have been frequently confounded with each other. The principal difference conſiſts in the beak, which, in this kind, is conical, very thick at the baſe, and tapering to a ſharp point: In this reſpect it more nearly reſembles the Groſbeak. Of this tribe many are diſtinguiſhed as well for the livelineſs of their ſong as for the beauty and variety of their plumage, on which accounts they are much eſteemed: They are very numerous, and aſſemble ſometimes in immenſe flocks, feeding on ſeeds and grain of various kinds, as well as inſects and their eggs.

THE HOUSE SPARROW.

(*Fringilla domeſtica*, Lin.—*Le Moineau franc*. Buff.)

The length of this bird is five inches and three quarters: The bill is duſky; eyes hazel; the top of the head and back part of the neck are aſh colour; the throat, fore part of the neck, and ſpace round the eyes, black; the cheeks are whitiſh; the breaſt and all the under parts are of a pale aſh colour; the back, ſcapulars, and wing coverts are of a reddiſh brown, mixed with black—the latter is tipped with white, forming a light bar acroſs the wing; the quills are duſky, with reddiſh edges; the tail is brown, edged with grey, and a little forked; the legs are pale brown. The female is diſtinguiſhed from the male in wanting the black patch on the throat, and in having a light ſtreak behind each eye; ſhe is alſo much plainer and duller in her whole plumage. In whatever country the Sparrow is ſettled, it is never found in deſert places, or at a diſtance from the dwellings of man: It does not, like other birds, ſhelter itſelf in woods and foreſts, or ſeek its ſubſiſtence in uninhabited plains, but is a reſident in towns and villages; it follows ſociety, and lives at its expence; granaries, barns, court-yards, pigeon-houſes, and in ſhort all places where grain is ſcattered, are its favorite reſorts. It is ſurely ſaying too much of this poor proſcribed ſpecies to ſum up its character in the words of the Count de Buffon:—"It is ex-"tremely deſtructive, its plumage is entirely uſe-"leſs, its fleſh indifferent food, its notes grating to "the ear, and its familiarity and petulance diſguſt-"ing." But let us not condemn a whole ſpecies of animals becauſe, in ſome inſtances, we have found them troubleſome or inconvenient. Of this we are ſufficiently ſenſible; but the uſes to which they are ſubſervient, in the grand economical diſtribution of nature, we cannot ſo eaſily aſcertain. We have already obſerved* that, in the deſtruction of caterpillars, they are eminently ſerviceable to vegetation, and in this reſpect alone there is reaſon to ſuppoſe ſufficiently repay the deſtruction they may make in the produce of the garden or the field. The great table of nature is ſpread

* See introduction.

out alike to all, and is amply ſtored with every thing neceſſary for the ſupport of the various families of the earth; it is owing to the ſuperior induſtry of man that he is enabled to appropriate ſo large a portion of the beſt gifts of providence for his own ſubſiſtence and comfort; let him not then think it waſte, that, in ſome inſtances, creatures inferior to him in rank are permitted to partake with him, nor let him grudge them their ſcanty pittance; but, conſidering them only as the taſters of his full meal, let him endeavour to imitate their chearfulneſs, and lift up his heart in grateful effuſions to Him, "who filleth all things living with plenteouſneſs."

The Sparrow never leaves us, but is familiar to the eye at all times, even in the moſt crowded and buſy parts of a town: It builds its neſt under the eaves of houſes, in holes of walls, and often about churches; it is made of hay, careleſsly put together, and lined with feathers: The female lays five or ſix eggs, of a reddiſh white colour, ſpotted with brown; ſhe has generally three broods in the year, from whence the multiplication of the ſpecies muſt be immenſe. Though familiar, the Sparrow is ſaid to be a crafty bird, eaſily diſtinguiſhing the ſnares laid to entrap it. In autumn prodigious flocks of them are ſeen every where, both in town and country; they often mix with other birds, and not unfrequently partake with the Pigeons or the poultry, in ſpite of every precaution to prevent them. The Sparrow is ſubject to great varieties of plumage: In the Britiſh and Leverian Muſeums there are ſeveral white ones, with yellow eyes and bills, others more or leſs mixed with brown, and ſome entirely black: A pair of white Sparrows were ſent us by Mr Walter Trevelyan, of St. John's College, Cambridge.—This bird, as ſeen in large and ſmoaky towns, is generally ſooty and unpleaſing in its appearance; but, among barns and ſtack-yards, the cock bird exhibits a very great variety in his plumage, and is far from being the leaſt beautiful of our Britiſh Birds.

THE MOUNTAIN SPARROW.

(*Fringilla Montana*, Lin.—*Le Friquet*, Buff.)

THIS bird is ſomewhat leſs than the common Sparrow: The bill is black; eyes hazel; the crown of the head and hind part of the neck are of a cheſtnut colour; ſides of the head white; throat black; behind each eye there is a pretty large black ſpot; the upper parts of the body are of a ruſty brown, ſpotted with black; the breaſt and under parts duſky white; the quills are black, with reddiſh edges, as are alſo the greater coverts; the leſſer are bay, edged with black, and croſſed with two white bars; the tail is of a reddiſh brown, and even at the end; the legs are pale yellow.

This ſpecies is frequent in Yorkſhire, Lancaſhire, and alſo in Lincolnſhire; it differs from the Houſe Sparrow in making its neſt in trees and not in buildings; it has not been ſeen further north than the above-mentioned counties. Buffon ſays that it feeds on fruits, ſeeds, and inſects; it is a lively, active little bird, and, when it alights, has a variety of motions, whirling about and jerking its tail upwards and downwards, like the Wagtail. It is found in Italy, France, Germany, and Ruſſia, and is much more plentiful in many parts of the continent than in England.

THE CHAFFINCH.

SHILFA, SCOBBY, SKELLY, OR SHELL-APPLE.

(*Fringilla cælebs*, Lin.—*Le Pinçon*, Buff.)

The bill is of a pale blue, tipped with black; eyes hazel; the forehead black; the crown of the head, hind part, and ſides of the neck are of a blueiſh aſh colour; ſides of the head, throat, fore part of the neck, and breaſt are of a vinaceous red; belly, thighs, and vent white, ſlightly tinged with red; the back is of a reddiſh brown, changing to green on the rump; both greater and leſſer coverts are tipped with white, forming two pretty large bars acroſs the wing; the baſtard wing and quill feathers are black, edged with yellow; the tail, which is a little forked, is black, the outermoſt feather edged with white; the legs are brown. The female wants the red upon the breaſt; her plumage in general is not ſo vivid, and inclines to green; in other reſpects it is not much unlike the male.

This beautiful little bird is every where well known; it begins its ſhort and frequently-repeated warble very early in the ſpring, and continues till about the ſummer ſolſtice, after which it is no more heard. It is a lively bird, and perpetually in motion, and this circumſtance has given riſe to the proverb, "*as gay as a Chaffinch.*" Its neſt is conſtructed with much art, of ſmall fibres, roots, and moſs, and lined with wool, hair, and feathers; the female lays generally five or ſix eggs, of a pale reddiſh colour, ſprinkled with dark ſpots, principally at the larger end. The male is very aſſiduous in his attendance during the time of hatching, ſeldom ſtraying far from the place, and then only to procure food. Chaffinches ſubſiſt chiefly on ſmall ſeeds of various kinds, they likewiſe eat caterpillars and inſects, with which they alſo feed their young. They are ſeldom kept in cages, as their ſong poſſeſſes no variety, and they are not very apt in learning the notes of other birds. The males frequently maintain obſtinate combats, and fight till one of them is vanquiſhed and compelled to give way. In Sweden theſe birds perform a partial migration; the females collect in vaſt flocks the latter end of September, and, leaving their mates, ſpread themſelves through various parts of

Europe: The males continue in Sweden, and are again joined by their females, who return in great numbers, about the beginning of April, to their wonted haunts. With us, both males and females continue the whole year. Mr White, in his Hiſtory of Selborne, obſerves, that great flocks ſometimes appear in that neighbourhood about Chriſtmas, and that they are almoſt entirely hens. It is difficult to account for ſo ſingular a circumſtance as the parting of the two ſexes in this inſtance; we would ſuppoſe that the males, being more hardy and better able to endure the rigours of the northern winters, are content to remain in the country, and pick up ſuch fare as they can find, whilſt the females ſeek for ſubſiſtence in more temperate regions.

THE MOUNTAIN FINCH.

BRAMBLING.

(*Fringilla Montifringilla*, Lin.—*Le Pinçon d'Ardennes*, Buff.)

Length ſomewhat above ſix inches: Bill yellow, blackiſh at the tip; eyes hazel; the feathers on the head, neck, and back are black, edged with ruſty brown; ſides of the neck, juſt above the wings, blue aſh; rump white; the throat, fore part of the neck, and breaſt are of a pale orange; belly white; leſſer wing coverts pale reddiſh brown, edged with white; greater coverts black tipped with pale yellow; quills duſky, with pale yellowiſh edges; the tail is forked, the outermoſt feathers edged with white, the reſt black, with whitiſh edges; legs pale brown.

The Mountain Finch is a native of northern climates, from whence it ſpreads into various parts of Europe: It arrives in this country the latter end of ſummer, and is more frequent in the mountainous parts of our iſland.* Great flocks of them ſometimes come together, they fly very cloſe, and on that account great numbers of them are frequently killed at one ſhot. In France they are ſaid to appear ſometimes in ſuch immenſe numbers, that the ground where they rooſted has been covered with their dung for a conſiderable ſpace; and in one year they were ſo numerous, that more than ſix hundred dozen were killed each night during the greateſt part of the winter.† They are ſaid to build their neſts in fir trees, at a conſiderable height; it is compoſed of long moſs, and lined with hair, wool, and feathers; the female lays four or five eggs, white, ſpotted with yellow. The fleſh of the Mountain Finch, though bitter, is ſaid to be good to eat, and better than that of the Chaffinch, but its ſong is much inferior, and is only a diſagreeable kind of chirping. It feeds on ſeeds of various kinds, and is ſaid to be particularly fond of beech maſt.

* We have ſeen them on the Cumberland hills in the middle of Auguſt.

† Buffon.

THE GOLDFINCH.

GOLDSPINK, OR THISTLE-FINCH.

(*Fringilla Carduelis*, Lin.—*Le Chardonneret*, Buff.)

The bill is white, tipped with black; the forehead and chin are of a rich ſcarlet colour, which is divided by a line paſſing from each corner of the bill to the eyes, which are black; the cheeks are white; top of the head black, which extends downward on each ſide, dividing the white on the cheeks, from the white ſpot on the hind part of the head; the back, rump, and breaſt are of a pale brown colour; belly white; greater wing coverts black; quills black, marked in the middle of each feather with yellow, forming, when the wing is cloſed, a large patch of that colour on the wing; the tips white; the tail feathers are black, with a white ſpot on each near the end; the legs are of a pale fleſh colour.

Beauty of plumage, ſays the lively Count de Buffon, melody of ſong, ſagacity, and docility of diſpoſition, ſeem all united in this charming little bird, which, were it rare, and imported from a foreign country, would be more highly valued. Goldfinches begin to ſing early in the ſpring, and continue till the time of breeding is over; when kept in a cage they will ſing the greateſt part of the year. In a ſtate of confinement they are much attached to their keepers, and will learn a variety of little tricks, ſuch as to draw up ſmall buckets containing their water and food, to fire a cracker, and ſuch like. They conſtruct a very neat and compact neſt, which is compoſed of moſs, dried graſs, and roots, lined with wool, hair, and the down of thiſtles, and other ſoft and delicate ſubſtances. The female lays five white eggs, marked with ſpots of a deep purple colour at the larger end: They feed their young with caterpillars and inſects; the old birds feed on various kinds of ſeeds, particularly the thiſtle, of which they are extremely fond. —Goldfinches breed with the Canary; this intermixture ſucceeds beſt between the cock Goldfinch and the hen Canary, whoſe offspring are productive, and are ſaid to reſemble the male in the ſhape of the bill, in the colours of the head and wings, and the hen in the reſt of the body.

THE SISKIN.

ABERDEVINE.

(*Fringilla Spinus*, Lin.—*Le Tarin*, Buff.)

Length near five inches: Bill white; eyes black; top of the head and throat black; over each eye there is a pale yellow ſtreak; back of the neck and back yellowiſh olive, faintly marked with duſky ſtreaks down the middle of each feather; rump yellow; under parts greeniſh yellow, paleſt on the breaſt; thighs grey, marked with duſky ſtreaks; greater wing coverts of a pale yellowiſh green, and tipped with black; quills duſky, faintly edged with yellow—the outer web of each at the baſe is of a fine pale yellow, forming, when the wing is cloſed, an irregular bar of that colour acroſs the wing; the tail is forked, the middle feathers black, with faint edges, the outer ones yellow, with black tips; the legs pale brown; claws white.

We have given the figure and defcription from one which we have kept many years in a cage; its fong, though not fo loud as the Canary, is pleafing and fweetly various; it imitates the notes of other birds, even to the chirping of the Sparrow: It is familiar, docile, and chearful, and begins its fong early in the mornings. Like the Goldfinch, it may eafily be taught to draw up its little bucket with water and food. Its food confifts chiefly of feeds; it drinks frequently, and feems fond of throwing water over its feathers: It breeds freely with the Canary. When a Sifkin is paired with the hen Canary, he is affiduous in his attention to his mate, carrying materials for the neft, and arranging them; and, during the time of incubation, regularly fupplying the female with food. Thefe birds are common in various parts of Europe; they are in moft places migratory, but do not feem to obferve any regular periods, as they are fometimes feen in large and at other times in very fmall numbers. Buffon obferves that thofe immenfe flights happen only once in the courfe of three or four years. It conceals its neft with fo much art, that it is extremely difficult to difcover it. Kramer obferves, that in the forefts bordering on the Danube thoufands of young Sifkins are frequently found, which have not dropt their firft feathers, and yet it is rare to meet with a neſt. It is not known to breed in this iſland, nor is it ſaid from whence they come over to us. Ours was caught upon the banks of the Tyne. In ſome parts of the South it is called the Barley-bird, being ſeen about that ſeed time; and in the neighbourhood of London it is known by the name of the Aberdevine.

CANARY FINCH.

(*Fringilla Canaria*, Lin.—*Le Serin des Canaries*, Buff.)

Is ſomewhat larger than the laſt, being about five inches and a half in length: The bill is of a pale fleſh colour; general colour of the plumage yellow, more or leſs mixed with grey, and in ſome with brown on the upper parts; the tail is ſomewhat forked; legs pale fleſh colour.

In a wild ſtate they are found chiefly in the Canary iſlands, from whence they have been brought to this country, and almoſt every part of Europe; they are kept in a ſtate of captivity, and partake of all the varieties attendant on that ſtate. Buffon enumerates twenty nine varieties, and many more might probably be added to the liſt, were all the changes incident to a ſtate of domeſtication carefully noted and brought into the account.—The breeding and rearing of theſe charming birds forms an amuſement of the moſt pleaſing kind, and affords a variety of ſcenes highly intereſting and gratifying to innocent minds. In the places fitted up and accommodated to the uſe of the little captives, we are delighted to ſee the workings of nature exemplified in the choice of their mates, building their neſts, hatching and rearing their young, and in the impaſſionate ardour exhibited by the male, whether he is engaged in aſſiſting his faithful mate in collecting materials for her neſt, in arranging them for her accommodation, in providing food for her offspring, or in chaunting his lively and amorous ſongs during every part of the important buſineſs. The Canary will breed freely with the Siſkin and Goldfinch, particularly the former, as we have already obſerved; it likewiſe proves prolific with the Linnet, but not ſo readily; and admits alſo the Chaffinch, Yellow Bunting, and even the Sparrow, though with ſtill more difficulty. In all theſe inſtances, except the firſt, the pairing ſucceeds beſt when the female Canary is introduced to the male of the oppoſite ſpecies. According to Buffon, the Siſkin is the only bird of which the male and female propagate equally with thoſe of the male or female Canaries.

The laſt-mentioned author, in his Hiſtory of Birds, has given a curious account of the various methods uſed in rearing theſe birds, to which we muſt refer our readers. We have thought it neceſſary to ſay thus much of a bird, which, though neither of Britiſh origin, nor yet a voluntary viſitor, muſt yet be conſidered as ours by adoption.* There are two kinds mentioned by Buffon, ſimilar to the Canary, both of them ſmaller; the former is called the Serin, the latter the Venturon, or Citril; they are both found in Italy, Greece, Turkey, and in the ſouthern provinces of France; they breed with the Canary, and are almoſt as remarkable for the ſweetneſs of their ſong.

THE LINNET.

GREY LINNET.

(*Fringilla Linaria*, Lin.—*La Linotte*, Buff.)

Length about five inches and a half: The bill blueiſh grey; eyes hazel; the upper parts of the head, neck, and back, are of a dark reddiſh brown, the edges of the feathers pale; the under parts are of a dirty reddiſh white; the breaſt is deeper than the reſt, and in ſpring becomes of a very beautiful crimſon; the ſides are ſpotted with brown; the quills are duſky, edged with white; the tail brown, likewiſe with white edges, except the two middle

* The importation of Canaries forms a ſmall article of commerce; great numbers are every year imported from Tyrol: Four Tyroleſe uſually bring over to England about ſixteen hundred of theſe birds; and though they carry them on their backs one thouſand miles, and pay twenty pounds for ſuch a number, they are enabled to ſell them at five ſhillings a piece.—*Phil. Tranſ. vol.* 62.

feathers, which have reddifh margins; it is fome-what forked; the legs are brown: The female wants the red on the breaft, inftead of which it is marked with ftreaks of brown; fhe has lefs white on her wings, and her colours in general are lefs bright.

This bird is very well known, being common in every part of Europe: it builds its neft in low bufhes; the outfide is made up of dried grafs, roots, and mofs; within it is lined with hair and wool: The female lays four or five eggs, of a pale blue colour, fpotted with brown at the larger end. She breeds generally twice in the year. The fong of the Linnet is beautiful and fweetly varied; its manners are gentle, and its difpofition docile; it eafily adopts the fongs of other birds, when confined with them, and in fome inftances has been faid to pronounce words with great diftinctnefs. This we confider as a perverfion of its talents, and fubftituting imperfect and forced accents, which have neither charms nor beauty, in the room of the free and varied modulations of uninftructed nature. Linnets are frequently found in flocks; during winter, they feed on various forts of feeds, and are faid to be particularly fond of lintfeed, from whence they derive their name.

THE GREATER REDPOLE.

(*Fringilla Cannabina,* Lin.—*Le grande Linotte de Vignes,* Buff.)

This bird is ſomewhat leſs than the laſt, and differs principally from the Linnet in being marked on the forehead by a blood-coloured ſpot; the breaſt likewiſe is tinged with a fine roſe colour; in other reſpects it reſembles the Linnet ſo much, that Buffon ſuppoſes them to be the ſame, and that the red ſpots on the head and breaſt are equivocal marks, differing at different periods, appearing at one time and diſappearing at another, in the ſame bird. It is certain that, during a ſtate of captivity, the red marks diſappear entirely; and that, in the time of moulting, they are nearly obliterated, and for ſome time do not recover their uſual luſtre. But hower plauſible this may appear, it is not well founded. The Redpole is ſmaller than the Linnet; it makes its neſt on the ground, while the latter builds in furze and thorn hedges: They differ likewiſe in the colour of their eggs—that of the Redpole being of a very pale green, with ruſty coloured ſpots: The head of the female is aſh-coloured, ſpotted with black, and of a dull yellow on the breaſt and ſides, which are ſtreaked with duſky lines.—Redpoles are common in the northern parts of England, where they breed chiefly in mountainous places.

LESSER REDPOLE.

(*Fringilla Linaria*, Lin.—*Le Sizerin*, Buff.)

Length about five inches: Bill pale brown, point duſky; eyes hazel; the forehead is marked with a pretty large ſpot, of a deep purpliſh red; the breaſt is of the ſame colour, but leſs bright; the feathers on the back are duſky, edged with pale brown; the greater and leſſer coverts tipped with dirty white, forming two light bars acroſs the wing; the belly and thighs are of a dull white; the quills and tail duſky, edged with dirty white; the latter ſomewhat forked; legs duſky. In our bird the rump was ſomewhat reddiſh, in which it agrees with the Twite of Mr Pennant, and moſt probably conſtitutes one ſpecies with it and the Mountain Linnet, the differences being immaterial, and merely ſuch as might ariſe from age, food, or other accidental circumſtances. The female has no red on the breaſt or rump, and the ſpot on her forehead is of a ſaffron colour; her plumage in general is not ſo bright as that of the male.

Birds of this kind are not unfrequent in this iſland; they breed chiefly in the northern parts, where they are known by the name of French Linnets. They make a ſhallow open neſt, compoſed of dried graſs and wool, and lined with hair and feathers: The female lays four eggs, almoſt white, marked with reddiſh ſpots. In the winter they mix with other birds, and migrate in flocks to the ſouthern counties: They feed on ſmall ſeeds of various kinds, eſpecially thoſe of the alder, of which they are extremely fond; they hang, like the Titmouſe, with their back downwards, upon the branches while feeding, and in this ſituation may eaſily be caught with lime twigs. This ſpecies is found in every part of Europe, from Italy to the moſt extreme parts of the Ruſſian empire. In America and the northern parts of Aſia it is likewiſe very common.

OF THE LARK.

Amongst the various kinds of ſinging birds with which this country abounds, there is none more eminently conſpicuous than thoſe of the Lark kind. Inſtead of retiring to woods and deep receſſes, or lurking in thickets, where it may be heard without being ſeen, the Lark is ſeen abroad in the fields; it is the only bird which chaunts on the wing, and as it ſoars beyond the reach of our ſight, pours forth the moſt melodious ſtrains, which may be diſtinctly heard at that amazing diſtance.—The great poet of nature thus beautifully deſcribes it as the leader of the general chorus:

———————— "Up ſprings the Lark,
"Shrill-voiced and loud, the meſſenger of morn;
"'Ere yet the ſhadows fly, he mounted ſings
"Amid the dawning clouds, and from their haunts
"Calls up the tuneful nations."

From the peculiar conſtruction of the hind claws, which are very long and ſtraight, Larks generally reſt upon the ground; thoſe which frequent trees perch only on the larger branches: They all build their neſts upon the ground, which expoſes them to the depredations of the ſmaller voracious kinds of animals, ſuch as the Weazel, Stoat, &c. which deſtroy great numbers of them. The Cuckoo likewiſe, which makes no neſt of its own, frequently ſubſtitutes its eggs in the place of theirs.—The general characters of this ſpecies are thus deſcribed :—The bill is ſtraight and ſlender, bending a little towards the end, which is ſharp-pointed ; the noſtrils are covered with feathers and briſtles ; the tongue is cloven at the end ; tail ſomewhat forked ; the toes divided to the origin—claw of the hind toe very long, and almoſt ſtraight ; the fore claws very ſhort, and ſlightly curved.

THE SKYLARK.

LAVROCK.

(*Alauda arvenſis*, Lin.—*L'Alouette*, Buff.)

Length near ſeven inches: Bill duſky, under mandible ſomewhat yellow; eyes hazel; over each eye there is a pale ſtreak, which extends to the bill, and round the eye on the under ſide; on the upper parts of the body the feathers are of a reddiſh brown colour, dark in the middle, with pale edges; the fore part of the neck is of a reddiſh white, daſhed with brown; breaſt, belly, and thighs white; the quills brown, with pale edges; tail the ſame, and ſomewhat forked, the two middle feathers darkeſt, the outermoſt white on the outer edge; the legs duſky. In ſome of our ſpecimens the feathers on the top of the head were long, and formed a ſort of creſt behind. The Leſſer Creſted Lark of Pennant and Latham is perhaps only a variety of this; the difference being trifling. It is ſaid to be found in Yorkſhire.

The Lark commences its ſong early in the ſpring, and is heard moſt in the morning: It riſes in the air almoſt perpendicularly and by ſucceſſive ſprings, and hovers at a vaſt height; its deſcent, on the contrary, is in an oblique direction, unleſs it is threatened by birds of prey, or attracted by its mate, and on theſe occaſions it drops like a ſtone. It makes its neſt on the ground, between two clods of earth, and lines it with dried graſs and roots; the female lays four or five eggs, of a greyiſh brown colour, marked with darker ſpots; ſhe generally has two broods in the year, and ſits only about fifteen days; as ſoon as the young have eſcaped from the ſhell, the attachment of the parent bird ſeems to increaſe; ſhe flutters over their heads, directs all their motions, and is ever ready to ſcreen them from danger. The Lark is almoſt univerſally diffuſed throughout Europe; it is every where extremely prolific, and in ſome places the prodigious numbers that are frequently caught are truly aſtoniſhing. In Germany there is an exciſe upon them, which has produced, according to Keyſler, the ſum of 6000 dollars yearly to the city of Leipſic alone. Mr Pennant ſays, the neighbourhood of Dunſtable is famous for the great numbers of theſe birds found there, and that 4000 dozen have been taken between September and February for the London markets. Yet, notwithſtanding the great havock made amongſt theſe birds, they are extremely numerous. The winter is the beſt ſeaſon for taking them, as they are then very fat, being almoſt conſtantly on the ground, feeding in great flocks; whereas in ſummer they are very lean; they then always go in pairs, eat ſparingly, and ſing inceſſantly while on the wing.

THE FIELD LARK.

(*Alauda campeſtris*, Lin.—*La Spipolette*, Buff.)

This exceeds the Titlark in ſize, being about ſix inches long: Its bill is ſlender; the plumage on the head, neck, and back is of a dark greeniſh brown, ſtreaked with black, paleſt on the rump; above each eye is a pale ſtreak: quill feathers duſky brown, with pale edges; the ſcapulars faintly bordered with white; the throat and under parts of the body are of a dirty white; the breaſt is yellowiſh, and marked with large black ſpots; the ſides and thighs ſtreaked with black; the tail duſky, two outer feathers white, excepting a ſmall part of the inner web, the two next tipped with white; the legs are of a yellowiſh brown; the hind claws ſomewhat curved.

Though much larger than the Titlark, this bird is ſimilar to it in plumage; its ſong is however totally different, as are alſo its haunts, being found chiefly near woods, and not unfrequently on trees; it builds its neſt like the laſt, and in ſimilar ſituations, on the ground, and ſometimes in a low buſh near the ground. The male is ſcarcely to be diſtinguiſhed from the female in its outward appearance. We have occaſionally met with another bird of the Lark kind, which we have ventured to denominate the Tree Lark; it frequents woods, and ſits on the higheſt branches of trees, from whence it riſes ſinging to a conſiderable height, deſcending ſlowly, with its wings and tail ſpread out like a fan. Its note is full, clear, melodious, and peculiar to its kind.

THE GRASHOPPER LARK.

(*Alauda trivialis*, Lin.—*L'Alouette Pipi*, Buff.)

This is the ſmalleſt of the Lark kind, and has, though we think not with ſufficient reaſon, been ranked among the warblers: Its bill is ſlender and duſky; the upper parts of the body are of a greeniſh colour, variegated and mixed with brown; the under of a yellowiſh white, ſpeckled irregularly on the breaſt and neck; the feathers of the wings and tail are of a paliſh duſky brown, with light edges;

the legs pale duſky brown; its hind claws, though ſhorter and more crooked than thoſe of the Sky-lark, ſufficiently mark its kind: It builds its neſt on the ground, in ſolitary ſpots, and conceals it beneath a turf; the female lays five eggs, marked with brown near the larger end.

In the ſpring the cock-bird ſometimes perches on a tall branch, ſinging with much emotion: At intervals he riſes to a conſiderable height, hovers a few ſeconds, and drops almoſt on the ſame ſpot, continuing to ſing all the time; his tones are ſoft, clear, and harmonious. In the winter its cry is ſaid to reſemble that of the graſhopper, but is rather ſtronger and ſhriller: It has been called the Pipit Lark from its ſmall ſhrill cry, and in German Piep-lerche for the ſame reaſon. Mr White obſerves, that its note ſeems cloſe to a perſon, though at an hundred yards diſtance; and when cloſe to the ear, ſeems ſcarce louder than when a great way off: It ſkulks in hedges and thick buſhes, and runs like a mouſe through the bottom of the thorns, evading the ſight. Sometimes, early in a morning, when undiſturbed, it ſings on the top of a twig, gaping and ſhivering with its wings.

THE WOODLARK.

(*Alauda arborea*, Lin.—*L'Alouette de bois*, Buff.)

This is ſomewhat ſmaller than the Field Lark, but reſembles it ſo much in the colours of its plumage as ſcarcely to need a ſeparate deſcription; in general they are much paler and leſs diſtinct; the ſtreak over each eye extends backwards towards the head, ſo as to form a ſort of wreath or coronet round it, which is very conſpicuous; the ſpots on its breaſt are larger and more diſtinct than thoſe of the Skylark, and its tail much ſhorter; the legs are of a dull yellow; the hind claw very long, and ſomewhat curved.

The Woodlark is generally found near the

borders of woods, from whence it derives its name; it perches on trees, and ſings during the night, ſo as ſometimes to be miſtaken for the Nightingale; it likewiſe ſings as it flies, and builds its neſt on the ground, ſimilar to that of the Skylark; the female lays five eggs, of a duſky hue, marked with brown ſpots: It builds very early, the young, in ſome ſeaſons, being able to fly about the latter end of March: She makes two neſts in the year, like the Skylark, but is not near ſo numerous as that bird. In autumn the Woodlarks are fat, and are then eſteemed excellent eating.

THE TITLARK.

(*Alauda pratenſis*, Lin.—*La Farlouſe ou L'Alouette, de prez*, Buff.)

This bird is leſs than the Woodlark, being not more than five inches and a half in length: Its bill is black at the tip, and of a yellowiſh brown at the baſe; its eyes are hazel; over each eye is a pale ſtreak; the diſpoſition of its colours is very ſimilar to thoſe of the Skylark, but ſomewhat darker on the upper parts, and inclining to a greeniſh brown; the breaſt is beautifully ſpotted with black on a light yellowiſh ground; the belly light aſh colour, obſcurely ſtreaked on the ſides with duſky; the tail is almoſt black, the two outer feathers white on the exterior edges, the outermoſt but one tipped with a white ſpot on the end; the legs are yellowiſh; feet and claws brown: The female differs only in its plumage being leſs bright than that of the male.

The Titlark is common in this country; and, though it ſometimes perches on trees, is generally found in meadows and low marſhy grounds:—It makes its neſt on the ground, lining it with hair; the female lays five or ſix eggs, of a deep brown colour; the young are hatched about the beginning of June. During the time of incubation the male ſits on a neighbouring tree, riſing at times and ſinging. The Titlark is fluſhed with the leaſt noiſe, and ſhoots with a rapid flight. Its note is fine, but ſhort, and without much variety; it warbles in the air like the Skylark, and increaſes its ſong as it deſcends ſlowly to the branch on which it chuſes to perch. It is further diſtinguiſhed by the ſhake of its tail, particularly whilſt it eats.

OF THE WAGTAIL.

THE different ſpecies of this kind are few, and theſe are chiefly confined to the continent of Europe, where they are very numerous. They are eaſily diſtinguiſhed by their briſk and lively motions, as well as by the great length of their tails, which they jerk up and down inceſſantly—from whence they derive their name.* They do not hop, but run along the ground very nimbly, after flies and other inſects, on which they feed: They likewiſe feed on ſmall worms, in ſearch of which they are frequently ſeen to flutter round the huſbandman whilſt at his plough, and follow the flocks in ſearch of the flies which generally ſurround them. They frequent the ſides of pools, and pick up the inſects which ſwarm on the ſurface. They ſeldom perch; their flight is weak and undulating, and during which they make a twittering noiſe.

* In almoſt all languages the name of this bird is deſcriptive of its peculiar habits. In Latin, Motacilla; in French, Motteux, La Lavandiere, or Waſher; in England, they are ſometimes called Waſhers, from their peculiar motion; in German, Brook-ſtilts; in Italian, Shake-tail, &c. &c.

THE PIED WAGTAIL.

BLACK AND WHITE WATER-WAGTAIL.

(*Motacilla Alba*, Lin.—*La Lavandiere*, Buff.)

The length of this bird is about ſeven inches: The bill is black; eyes hazel; hind part of the head and neck black; the forehead, cheeks, and ſides of the neck are white; the fore part of the neck and part of the breaſt are black, bordered by a line of white, in the form of a gorget; the back and rump are of a deep aſh colour; wing coverts and ſecondary quills duſky, edged with light grey; prime quills black, with pale edges; lower part of the breaſt and belly white; the middle feathers of the tail are black, the outermoſt white, except at the baſe and tips of the inner webs, which are black; legs black. There are ſlight variations in theſe birds; ſome are white on the chin and throat, leaving only a creſcent of black on the breaſt. The head of the female is brown.

This is a very common bird with us, and may be ſeen every where, running on the ground, and frequently leaping after flies and other inſects, on which it feeds. Its uſual haunts are the ſhallow margins of waters, into which it will ſometimes wade a little in ſearch of its food. It makes its neſt on the ground, of dry graſs, moſs, and ſmall roots, lined with hair and feathers; the female lays five white eggs, ſpotted with brown. The parent birds are very attentive to their young, and continue to feed and train them for three or four weeks after they are able to fly; they will defend them with great courage when in danger, or endeavour to draw aſide the enemy by various little arts. They are very attentive to the cleanlineſs of the neſt, and will throw out the excrement; they have been known to remove light ſubſtances, ſuch as paper or ſtraw, which has been laid as a mark for the neſt. It is ſaid by ſome authors to migrate into other climates about the end of October; with us it is known to change its quarters as the winter approaches, from north to ſouth. Its note is ſmall and inſignificant, but frequently repeated, eſpecially while on the wing.

THE GREY WAGTAIL.

(*Motacilla Boarula*, Lin.—*La Bergeronette jaune*, Buff.)

This bird is ſomewhat larger than the laſt, owing to the great length of its tail: Its bill is dark brown; over each eye there is a pale ſtreak; the head, neck, and back are of a greyiſh aſh colour; the throat and chin are black; the rump and all the under parts of the body are of a bright yellow; wing coverts and quills dark brown, the former with pale edges; the ſecondaries, which are almoſt as long as the greater quills, are white at the baſe, and tipped with yellow on the outer edges; the middle feathers of the tail are black, the outer ones white; legs yellowiſh brown.

This elegant little bird frequents the ſame places as the laſt; its food is likewiſe ſimilar to it. It remains with us during winter, frequenting the neighbourhood of ſprings and running waters: The female builds her neſt on the ground, and ſometimes in the banks of rivulets; it is compoſed of nearly the ſame materials as the laſt; ſhe lays from ſix to eight eggs, of a dirty white, marked with yellow ſpots: She differs from the male in having no black on the throat.

THE YELLOW WAGTAIL.

(*Motacilla Flava*, Lin.—*La Bergeronette de printems*, Buff.)

Length ſix inches and a half: Bill black; eyes hazel; the head and all the upper parts of the body are of an olive green, paleſt on the rump; the under parts are of a bright yellow, dafhed with a few dark fpots on the breaft and belly; over each eye there is a pale yellow ftreak, and beneath a dufky line, curving upwards towards the hind part of the head; wing coverts edged with pale yellow; quills dufky; tail black, except the outer feathers, which are white; the legs are black; hind claws very long.

Buffon obferves that this bird is feen very early in the fpring, in the meadows and fields, amongft the green corn, where it frequently neftles; it haunts the fides of brooks and fprings which never freeze with us during winter. The female lays five eggs, of a pale lead colour, with dufky fpots irregularly difpofed.

OF THE FLYCATCHERS.

Of thofe birds which conftitute this clafs we only find two kinds which inhabit this ifland, and thefe are not the moft numerous of the various tribes with which this country abounds. The ufeful inftincts and propenfities of this little active race are chiefly confined to countries under the more immediate influence of the fun, where they are of infinite ufe in deftroying thofe numerous fwarms of noxious infects engendered by heat and moifture, which are continually upon the wing. Thefe, though weak and contemptible when individually confidered are formidable by their numbers, devouring the whole produce of vegetation, and carrying in their train the accumulated ills of peftilence and famine. Thus, to ufe the words of an eminent Naturalift,* " we fee, that all nature is balanced, and the circle of generation and deftruction is perpetual! The philofopher contemplates with melancholy this feemingly cruel fyftem, and ftrives in vain to reconcile it with his ideas of benevolence, but he is forcibly ftruck with the nice adjuftment of the various parts, their mutual connection and fubordination, and the unity of plan which pervades the whole."

* Buffon.

The characters of this genus with us are somewhat equivocal and not well aſcertained, neither do we know of any common name in our language by which it is diſtinguiſhed. Mr Pennant deſcribes it thus: "Bill flatted at the baſe, almoſt triangular, notched at the end of the upper mandible, and beſet with briſtles at its baſe." We have placed the Flycatcher here, as introductory to the numerous claſs which follows, to which they are nearly related, both in reſpect to form, habits, and modes of living: The affinity between them is ſo great, as to occaſion ſome confuſion in the arrangement of ſeveral of the individuals of each kind, for which reaſon we have placed them together.

THE PIED FLYCATCHER.

COLDFINCH.

(*Muſcicapa Atricapilla*, Lin.—*Le traquet d'Angleterre*, Buff.)

Length near five inches: Bill black; eyes hazel; the forehead is white; the top of the head, back, and tail are black; the rump is daſhed with aſh colour; the wing coverts are duſky, the greater coverts are tipped with white; the exterior ſides of the ſecondary quills are white, as are alſo the outer feathers of the tail; all the under parts, from the bill to the tail, are white; the legs are black: The female is brown where the male is black; it likewiſe wants the white ſpot on the forehead. This bird is no where common; it is in moſt plenty in Yorkſhire, Lancaſhire, and Derbyſhire. Since the cut, which was done from a ſtuffed ſpecimen, was finiſhed, we have been favoured with a pair of theſe birds, ſhot at Benton, in Northumberland: We ſuppoſe them to be male and female, as one of them wanted the white ſpot on the forehead; in other reſpects it was ſimilar to the male: The upper parts in both were black, obſcurely mixed with brown; the quill feathers dark reddiſh brown; tail dark brown, the exterior edge of the outer feather white; legs black.

SPOTTED FLYCATCHER.

BEAM-BIRD.

(*Muſcicapa Griſola*, Lin.—*Le Gobe-mouche*, Buff.)

Length near five inches and three quarters: Bill duſky, baſe of it whitiſh, and beſet with ſhort briſtles; inſide of the mouth yellow; the head and back light brown, obſcurely ſpotted with black; the wings duſky, edged with white; the breaſt and belly white; the throat and ſides under the wings tinged with red; the tail duſky; legs black.

Mr White obſerves, that the Flycatcher, of all our ſummer birds, is the moſt mute and the moſt familiar. It viſits this iſland in the ſpring, and diſappears in September; it builds in a vine or ſweet-briar, againſt the wall of a houſe, or on the end of a beam, and ſometimes cloſe to the poſt of a door where people are going in and out all day long; it returns to the ſame place year after year: The female lays four or five eggs, marked with ſmall ruſty ſpots; the neſt is careleſsly made, and conſiſts chiefly of moſs, frequently mixed with wool and ſtrong fibres, ſo large, ſays Buffon, that it appears ſurprizing how ſo ſmall an artificer could make uſe of ſuch ſtubborn materials. This bird feeds on inſects, which it catches on the wing; it ſometimes watches for its prey, ſitting on a branch or poſt, and, with a ſudden ſpring, takes it as it flies, and immediately returns to its ſtation to wait for more; it is likewiſe fond of cherries. Mr Latham ſays, it is known in Kent by the name of the Cherry-ſucker. It has no ſong, but only a ſort of inward wailing note, when it perceives any danger to itſelf or young: It breeds only once, and retires early. When its young are able to fly, it retires with them to the woods, where it ſports with them among the higher branches, ſinking and riſing often perpendicularly among the flies which hum below.

OF THE WARBLERS.

THIS very numerous claſs is compoſed of a great variety of kinds, differing in ſize from the Nightingale to the Wren, and not a little in their habits and manners. They are widely diſperſed over moſt parts of the known world; ſome of them remain with us during the whole year—others are migratory, and viſit us annually in great numbers, forming a very conſiderable portion of thoſe numerous tribes of ſinging birds, with which this iſland ſo plentifully abounds. Some of them are diſtinguiſhed by their manner of flying, which they perform by jerks, and in an undulating manner; others by the whirring motion of their wings. The head in general is ſmall; the bill is weak and ſlender, and beſet with briſtles at the baſe; the noſtrils are ſmall and ſomewhat depreſſed; and the outer toe is joined to the middle one by a ſmall membrane.

THE NIGHTINGALE.

(*Motacilla luſcinia*, Lin.—*Le Roſſignol*, Buff.)

This bird, ſo deſervedly eſteemed for the excellence of its ſong, is not remarkable for the variety or richneſs of its colours; it is ſomewhat more than ſix inches in length: Its bill is brown, yellow on the edges at the baſe; eyes hazel; the whole upper part of the body is of a ruſty brown, tinged with olive; the under parts pale aſh colour, almoſt white at the throat and vent; the quills are brown, with reddiſh margins; legs pale brown. The male and female are very ſimilar.

Although the Nightingale is common in this country, it never viſits the northern parts of our iſland, and is but ſeldom ſeen in the weſtern counties of Devonſhire and Cornwall: It leaves us ſome time in the month of Auguſt, and makes its regular return the beginning of April; it is ſuppoſed, during that interval, to viſit the diſtant regions of Aſia; this is probable, as they do not winter in any part of France, Germany, Italy, Greece, &c. neither does it appear that they ſtay in Africa, but are ſeen at all times in India, Perſia, China, and Japan; in the latter place they are much eſteemed for their ſong, and ſell at great prices. They are ſpread generally throughout Europe, even as far north as Siberia and Sweden, where they are ſaid to ſing delightfully; they, however, are partial to particular places, and avoid others which ſeem as likely to afford them the neceſſary means of ſupport. It is not improbable, however, that, by planting a colony in a well-choſen ſituation, theſe delightful ſongſters might be induced to haunt places where they are not at preſent ſeen; the experiment might be eaſily tried, and, ſhould it ſucceed, the reward would be great in the rich and varied ſong of this unrivalled bird. The following animated deſcription of it is taken from the ingenious author of the *Hiſtoire des Oiſeaux*:—"The leader of the vernal chorus begins with a "low and timid voice, and he prepares for the "hymn to nature by eſſaying his powers and at-"tuning his organs; by degrees the ſound opens "and ſwells, it burſts with loud and vivid flaſhes, "it flows with ſmooth volubility, it faints and mur-

"murs, it ſhakes with rapid and violent articula- "tions; the ſoft breathings of love and joy are "poured from his inmoſt ſoul, and every heart "beats uniſon, and melts with delicious languor. "But this continued richneſs might ſatiate the ear. "The ſtrains are at times relieved by pauſes, "which beſtow dignity and elevation. The mild "ſilence of evening heightens the general effect, "and not a rival interrupts the ſolemn ſcene."— Theſe birds begin to build about the end of April or the beginning of May; they make their neſt in the lower part of a thick buſh or hedge; the female lays four or five eggs, of a greeniſh brown colour; the neſt is compoſed of dry graſs and leaves, intermixed with ſmall fibres, and lined with hair, down, and other ſoft and warm ſubſtances. The buſineſs of incubation is entirely performed by the female, whilſt the cock, at no great diſtance, entertains her with his delightful melody; ſo ſoon, however, as the young are hatched, he leaves off ſinging, and joins her in the care of providing for the young brood. Theſe birds make a ſecond hatch, and ſometimes a third; and in hot countries they are ſaid to have four.

The Nightingale is a ſolitary bird, and never unites in flocks like many of the ſmaller birds, but hides itſelf in the thickeſt parts of the buſhes, and ſings generally in the night: Its food conſiſts principally of inſects, ſmall worms, eggs of ants, and ſometimes berries of various kinds. Nightingales, though timorous and ſhy, are eaſily caught; ſnares of all ſorts are laid for them, and generally ſucceed; they are likewiſe caught on lime twigs:— Young ones are ſometimes brought up from the neſt, and fed with great care till they are able to ſing. It is with great difficulty that old birds are induced to ſing after being taken; for a conſiderable time they refuſe to eat, but by great attention to their treatment, and avoiding every thing that might agitate them, they at length reſume their ſong, and continue it during the greateſt part of the year.

THE DARTFORD WARBLER.

(*Le Pitchou de Provence*, Buff.)

This bird meaſures above five inches in length, of which the tail is about one half: Its bill is long and ſlender, and a little bent at the tip; it is of a black colour, whitiſh at the baſe; its eyes are reddiſh; eye-lids deep crimſon; all the upper parts are of a dark ruſty brown, tinged with dull yellow; the breaſt, part of the belly, and thighs are of a deep red, inclining to ruſt colour; the middle of the belly is white; the baſtard wing is alſo white; the tail is duſky, except the exterior web of the outer feather, which is white; the legs are yellow.

This ſeems to be a rare bird in this country, and owes its name, with us, to the accident of a pair of them having been ſeen near Dartford, in Kent, a few years ago; they have ſince been obſerved in greater numbers, and are ſuppoſed ſometimes to winter with us. Buffon ſays they are natives of Provence, where they frequent gardens, and feed on flies and ſmall inſects. Our repreſentation was taken from a ſtuffed ſpecimen in the Wycliffe Muſeum, now in the poſſeſſion of Geo. Allan, Eſq. of the Grange, near Darlington.

THE REDBREAST.

ROBIN-REDBREAST, OR RUDDOCK.

(*Motacilla rubecola*, Lin.—*Le Rouge-gorge*, Buff.)

This general favorite is too well known to need a very minute deſcription: Its bill is ſlender and delicate; its eyes are large, black, and expreſſive, and its aſpect mild; its head and all the upper parts of its body are brown, tinged with a greeniſh olive; its neck and breaſt are of a fine deep reddiſh orange; a ſpot of the ſame colour marks its forehead; its belly and vent are of a dull white; its legs are duſky.

During the ſummer the Redbreaſt is rarely to be ſeen; it retires to woods and thickets, where, with its mate, it prepares for the accommodation of its future family. Its neſt is placed near the ground, by the roots of trees, in the moſt concealed ſpot, and ſometimes in old buildings; it is conſtructed of moſs, intermixed with hair and dried leaves, and lined with feathers: In order more effectually to conceal it, the bird covers its neſt with leaves, leaving only a narrow winding entrance under the heap. The female lays from five to nine eggs, of a dull white, marked with reddiſh ſpots. During the time of incubation, the cock ſits at no great diſtance, and makes the woods reſound with his delightful warble; he keenly chaſes all the birds of his own ſpecies, and drives them from his little ſettlement; for, as faithful as they are amorous, it has never been obſerved that two pairs of theſe birds were ever lodged in the ſame buſh.* The Redbreaſt prefers the thick ſhade, where there is water; it feeds on inſects and worms; its delicacy in preparing the latter is ſomewhat remarkable:—It takes it by one end, in its beak, and beats it on the ground till the

* Unum arbuſtum non alit duos erithacos.

inward part comes away; then, taking it by the other in like manner, cleanses it from all its impurities, eating only the outward part or skin.—Although the Redbreast never quits this island, it performs a partial migration. As soon as the business of incubation is over, and the young are sufficiently grown to provide for themselves, it leaves its retirement, and again draws near the habitations of mankind: Its well-known familiarity has attracted the attention and secured the protection of men in all ages; it haunts the dwellings of the cottager, and partakes of his humble fare; when the cold grows severe, and snow covers the ground, it approaches the house, taps at the window with its bill, as if to entreat an asylum, which is always chearfully granted, and, with a simplicity the most delightful, hops round the house, picks up crumbs, and seems to make himself one of the family.—Thomson has very beautifully described the annual visits of this little guest in the following lines:

> The Redbreast, sacred to the household gods,
> Wisely regardful of th' embroiling sky,
> In joyless fields and thorny thickets leaves
> His shivering mates, and pays to trusted man
> His annual visit. Half afraid, he first
> Against the window beats; then brisk alights
> On the warm hearth; then, hopping o'er the floor,
> Eyes all the smiling family askance,
> And pecks, and starts, and wonders where he is;
> Till, more familiar grown, the table crumbs
> Attract his slender feet.

The young Redbreaſt, when full feathered, may be taken for a different bird, being ſpotted all over with ruſt-coloured ſpots on a light ground: The firſt appearance of the red is about the end of Auguſt, but it does not arrive at its full colour till the end of the following month. Redbreaſts are never ſeen in flocks, but always ſingly; and, when all other birds aſſociate together, they ſtill retain their ſolitary habits. Buffon ſays, that as ſoon as the young birds have attained their full plumage, they prepare for their departure; but in thus changing their ſituation, they do not gather in flocks, but perform their journey ſingly, one after another; which is a ſingular circumſtance in the hiſtory of this bird. Its general familiarity has occaſioned it to be diſtinguiſhed by a peculiar name in many countries: About Bornholm it is called Tomi Liden; in Norway, Peter Ronſmad; in Germany it is called Thomas Gierdet; and with us, Robin-Redbreaſt, or Ruddock.

THE REDSTART.

RED-TAIL.

(*Motacilla Phœnicurus*, Lin.—*Le Roſſignol de muraille*, Buff.)

This bird meaſures rather more than five inches in length: Its bill and eyes are black; its forehead is white; cheeks, throat, fore part and ſides of the neck black, which colour extends over each eye; the crown of its head, hind part of its neck, and back are of a deep blue grey; in ſome ſubjects, probably old ones, this grey is almoſt black; its breaſt, rump, and ſides are of a fine glowing red, inclining to orange colour, which extends to all the feathers of the tail, except the two middle ones, which are brown; the belly is white; feet and claws black. The female differs conſiderably from the male; the top of the head and back are of a grey aſh colour; the chin is white, and its colour not ſo vivid.

The Redſtart is migratory; it appears about the middle of April, and departs the latter end of September, or beginning of October; it frequents old walls and ruinous edifices, where it makes its neſt, compoſed chiefly of moſs, lined with hair and feathers: It is diſtinguiſhed by a peculiar quick ſhake of its tail from ſide to ſide on its alighting on a wall or other place. Though a wild and timorous bird, it is frequently found in the midſt of cities, always chuſing the moſt difficult and inacceſſible places for its reſidence; it likewiſe builds in foreſts, in holes of trees, or in high and dangerous precipices; the female lays four or five eggs, not much unlike thoſe of the Hedge-ſparrow, but ſomewhat longer. Theſe birds feed on flies, ſpiders, the eggs of ants, ſmall berries, ſoft fruits, and ſuch like.

THE FAUVETTE.

PETTICHAPS.

(*Motacilla hippolais*, Lin.—*La Fauvette*, Buff.)

Length about ſix inches: Its bill is blackiſh; eyes dark hazel; the whole upper part of the body is of a dark brown or mouſe colour, lightly tinged with pale brown on the edges of the wing coverts, and along the webs of the ſecondary quills; the larger quills are of a duſky aſh colour, as are alſo thoſe of the tail, except the outermoſt, which are white on their exterior ſides and tips; over each eye there is a pale ſtreak; the throat and belly are of a ſilvery white; legs dark brown.

This bird frequents thickets, and is ſeldom to be ſeen out of covert; it ſecretes itſelf in the thickeſt parts of the buſhes, from whence it may be heard, but not ſeen: It is truly a mocking bird, imitating the notes of various kinds, generally beginning with thoſe of the Swallow, and ending with the full ſong of the Blackbird. We have often watched with the utmoſt attention whilſt it was ſinging delightfully in the midſt of a buſh cloſe at hand, but have ſeldom been able to obtain a ſight of it: We could never procure more than one ſpecimen:—Its appearance with us does not ſeem to be regular, as we have frequently been diſappointed in not finding it in its uſual haunts. We ſuppoſe this to be the ſame with the Fauvette of M. Buffon,* which he places at the head of a numerous family, conſiſting of ten diſtinct ſpecies; many of which viſit this iſland in the ſpring, and leave it again in autumn. "Theſe pretty warblers," ſays he, "arrive when the trees put forth their leaves, and be-

* We have adopted the name of *Fauvette* for want of a more appropriate term in our own language. We apprehend this to be the *Flycatcher* of Mr Pennant—*Br. Zool. vol. 2d, p.* 264, *1ſt ed.*—and the *Leſſer Pettichaps* of Latham, which he ſays is known in Yorkſhire by the name of the Beam-bird; but he does not ſpeak from his own knowledge of the bird. It certainly is but little known, and has no common name in this country.

gin to expand their bloſſoms; they diſperſe through the whole extent of our plains; ſome inhabit our gardens, others prefer the clumps and avenues; ſome conceal themſelves among the reeds, and many retire to the midſt of the woods." But, notwithſtanding their numbers, this genus is confeſſedly the moſt obſcure and indetermined in the whole of ornithology. We have taken much pains to gain a competent knowledge of the various kinds which viſit our iſland, and have procured ſpecimens of moſt, if not all of them, but confeſs that we have been much puzzled in reconciling their provincial names with the ſynonima of the different authors who have noticed them.

The following is deſcribed by Latham as a variety of the Pettichaps, and agrees in moſt reſpects with our ſpecimen. We conceive it to be the ſame as the Paſſerinette of Buffon, allowing ſomewhat for difference of food, climate, &c.

THE LESSER FAUVETTE.

PASSERINE WARBLER.

(*Motacilla paſſerina*, Lin.—*Le Paſſerinette*, Buff.)

Length nearly the ſame as the laſt: Bill pale brown; upper parts of the body brown, ſlightly tinged with olive green; under parts duſky white, a little inclining to brown acroſs the breaſt; quills duſky, with pale edges; tail duſky; over each eye there is an indiſtinct whitiſh line; legs pale brown. The male and female are much alike: The eggs are of a dull white, irregularly marked with duſky and black ſpots.—This bird is alſo a mocker, but its ſong is not ſo powerful as the laſt.

THE WINTER FAUVETTE.

HEDGE WARBLER, HEDGE SPARROW, OR DUNNOCK.

(*Motacilla Modularis*, Lin.—*La Fauvette d'hiver*, Buff.)

The length of this well-known bird is ſomewhat more than five inches: Its bill is dark; eyes hazel; its general appearance is that of a duſky brown, moſt of the feathers on the back and wings being edged with reddiſh brown; the cheeks, throat, and fore part of its neck are of a dull blueiſh aſh colour; the belly is of a dirty white; quills and tail duſky; rump greeniſh brown; ſides and thighs pale tawny brown; the legs are brown.

This bird is frequently ſeen in hedges, from whence it derives one of its names; but it has no other relation to the Sparrow than in the dinginefs of its colours; in every other refpect it differs entirely. It remains with us the whole year, and builds its neft near the ground; it is compofed of mofs and wool, and lined with hair; the female generally lays four or five eggs, of a uniform pale blue, without any fpots: The young are hatched about the beginning of May. During the time of fitting, if a cat or other voracious animal fhould happen to come near the neft, the mother endeavours to divert it from the fpot by a ftratagem fimilar to that by which the Partridge mifleads the dog: She fprings up, flutters from fpot to fpot, and by that means allures her enemy to a fafe diftance. In France, the Hedge-fparrow is rarely feen but in winter; it arrives generally in October, and departs in the fpring for more northern regions, where it breeds. It is fuppofed to brave the rigours of winter in Sweden, and that it affumes the white plumage common in thofe fevere climates in that feafon. Its fong is little varied, but pleafant, efpecially in a feafon when all the other warblers are filent: Its ufual ftrain is a fort of quivering, frequently repeating fomething like the following *tit-tit-tititit*, from whence, in fome places, it is called the Titling. We have already obferved that the Cuckoo frequently makes ufe of the neft of this bird to depofit her egg in.

THE REED FAUVETTE.

SEDGE BIRD.

(*Motacilla Salicaria*, Lin.—*La Fauvette de roſeaux*, Buff.)

This elegant little bird is about the ſize of the Black-cap: Its bill is duſky: eyes hazel; the crown of the head and back are brown, marked with duſky ſtreaks; the rump tawny; the cheeks are brown; over each eye there is a light ſtreak; the wing coverts are duſky, edged with pale brown, as are alſo the quills and tail; the throat, breaſt, and belly are white—the latter tinged with yellow; the thighs are yellow; legs duſky; the hind claws are long and much bent.

This bird is found in places where reeds and ſedges grow, and builds its neſt there; it is made of dried graſs and tender fibres of plants, and lined with hair, and uſually contains five eggs, of a dirty white, mottled with brown ; it likewiſe frequents the ſides of rivers and ponds where there is covert : It ſings inceſſantly night and day, during the breeding time, imitating by turns the notes of the Sparrow, the Swallow, the Skylark, and other birds—from whence it is called the Engliſh Mock-bird. Buffon obſerves, that the young ones, though tender and not yet fledged, will deſert the neſt if it be touched, or even if a perſon go too near it. This diſpoſition, which is common to all the Fauvettes, as well as to this which breeds in watery places, ſeems to characteriſe the inſtinctive wildneſs of the whole ſpecies.

THE BLACK-CAP.

(*Motacilla Atricapilla*, Lin.—*La Fauvette à tête noire*, Buff.)

This bird is in length ſomewhat above five inches: The upper mandible is of a dark horn colour; the under one light blue—edges of both whitiſh; top of the head black; ſides of the head and back of the neck aſh colour; back and wings of an olive grey; the throat and breaſt are of a ſilvery grey; belly and vent white; the legs are of a blueiſh colour, inclining to brown; the claws black: The head of the female is of a dull ruſt colour.

The Blackcap viſits us about the middle of April, and retires in September; it frequents gardens, and builds its neſt near the ground, which is compoſed of dried graſs, moſs, and wool, and lined with hair and feathers; the female lays five eggs, of a pale reddiſh brown, ſprinkled with ſpots of a darker colour. During the time of incubation the male attends the female, and fits by turns; he likewife procures her food, fuch as flies, worms, and infects. The Black-cap fings fweetly, and fo like the Nightingale, that in Norfolk it is called the Mock-nightingale. Buffon fays that its airs are light and eafy, and confift of a fucceffion of modulations of fmall compafs, but fweet, flexible, and blended. And our ingenious countryman, Mr White, obferves, that it has ufually a full, fweet, deep, loud, and wild pipe, yet the ftrain is of fhort continuance, and its motions defultory; but when this bird fits calmly, and in earneft engages in fong, it pours forth very fweet but inward melody, and expreffes great variety of fweet and gentle modulations, fuperior perhaps to any of our warblers, the Nightingale excepted; and, while they warble, their throats are wonderfully diftended. Black-caps feed chiefly on flies and infects, and not unfrequently on ivy and other berries.

THE WHITE-THROAT.

MUGGY.

(*Motacilla ſylvia*, Lin.—*La Fauvette griſe*, Buff.)

THE length of this bird is about five inches and a half: Its bill is dark brown, lighter at the baſe; eyes dark hazel; the upper part of the head and back are of a reddiſh aſh colour; throat white; leſſer wing coverts pale brown; the greater duſky brown, with reddiſh margins; breaſt and belly ſilvery white; the wings and tail are duſky brown, with pale edges, the outer feathers white; the legs pale brown: The breaſt and belly of the female are entirely white.

This bird arrives with the Redſtart, Black-cap, &c. in the ſpring, and quits us in autumn about the ſame time with them; it frequents thickets and hedges, and feeds on infects and wild berries; it makes its neft in thick bufhes, of dried grafs and mofs; the female lays five eggs, of a greenifh white, fprinkled with dark fpots. Its note, which is rather harfh and unpleafing, is frequently repeated, and is attended with a particular motion of the wings; it is fhy and wild, and is not frequently found near the habitations of men.

THE YELLOW WILLOW WREN.

(*Motacilla trochilus*, Lin.—*Le Pouillot, ou le chantre*, Buff.)

Length nearly five inches: The bill is brown, the infide and edges yellow; eyes hazel; the upper parts of the plumage are yellow, inclining to a pale olive green; the under pale yellow; over each eye there is a whitifh ftreak, which in young birds we have obferved to be particularly diftinct; the wings and tail are of a dufky brown, with pale edges; legs yellowifh brown.

The ingenious Mr White obferves, that there are three diftinct fpecies of the Willow Wren, of which this is the largeft; the two following differ in their fize as well as note; their form and manners are however very fimilar: We have been fortunate in procuring fpecimens of each kind, taken at the fame time of the year, and had an opportunity of noticing the difference of their fong,

For ſpecimens of all the birds of this kind, as well as many others, we are indebted to Lieut. H. F. Gibſon, of the 4th dragoons, whoſe kind attention to our work merits our warmeſt acknowledgement.—This bird is frequent on the tops of trees, from whence it often riſes ſinging; its note is rather low, but ſoft, and ſweetly varied. It arrives in this country early in the ſpring, and departs in autumn; it makes its neſt in holes, at the roots of trees, or in dry banks; it is arched ſomewhat like that of the Wren, and is made chiefly of moſs, lined with wool and hair; the female lays from five to ſeven eggs, of a dirty white, marked with reddiſh ſpots.

THE WILLOW WREN.

(*Le Figuier brun et jaune*, Buff.)

This is next in ſize: Its bill is brown, the upper parts of a greeniſh olive colour, darker than the laſt; over each eye a light yellow line extends from the bill to the back part of the head; the wings are brown, with light yellowiſh edges; the throat and breaſt are white, with a pretty ſtrong tinge of yellow; the belly is whitiſh; thighs yellow; legs yellowiſh brown,—as is likewiſe the inſide of the bill. They vary much in colour.

We are favoured, by the ingenious Mr I. Gough of Kendal, with the deſcription of a bird very ſimilar to this, which is common in Weſtmoreland, where it is known by the name of the Strawſmeer. It appears in the vallies in April, a few days after the Swallow, and begins to ſing immediately on its arrival, and may be heard till the beginning of Auguſt; it frequents hedges, ſhrubberies, and ſuch like places; its food conſiſts of inſects, in ſearch of which it is continually running up and down ſmall branches of trees: It makes an artleſs neſt, of withered graſs, moſs, and the ſlender ſtems of dried plants, it is lined with feathers, hair, and a little wool, and is commonly placed in a low thick buſh or hedge; the female generally lays five eggs, of a dirty white, marked at the larger end with numerous dark brown oval ſpots. We ſuppoſe this to be the Scotch Warbler of Mr Pennant, and the Figuier brun et jaune of M. Buffon.

THE LEAST WILLOW WREN.

THE upper parts of the plumage of this bird are darker than the two laſt, ſomewhat inclining to a mouſe colour: Its breaſt is of a dull ſilvery white, from whence in ſome places it is called the Lintywhite; its legs are dark.

The ſong of this is not ſo loud as the laſt, though very ſimilar, and conſiſts of a ſingle ſtrain, very weak, and frequently repeated; they are both common in woods and coverts, warbling their little ſimple ſong as they ſit upon the branches of trees.

THE GOLDEN-CRESTED WREN.

(*Motacilla regulus*, Lin.—*Le Roitelet*, Buff.)

This is ſuppoſed to be the leaſt of all the European birds; it is certainly the ſmalleſt of the Britiſh kinds, being in length not quite three inches and a half,* and weighs only ſeventy-ſix grains: Its bill is very ſlender and dark; eyes hazel; on the top of its head the feathers are of a bright orange colour, bordered on each ſide with black, which forms an arch above its eyes, and with which it ſometimes conceals the crown, by contracting the muſcles of the head; the upper part of the body is of a yellowiſh green or olive colour; all the un-

* The body, when ſtripped of its feathers, is not quite an inch long.—*Buff.*

der parts are of a pale reddiſh white, tinged with green on the ſides; the greater coverts of the wings are of a duſky brown, edged with yellow, and tipped with white; quills duſky, edged with pale green, as are alſo the feathers of the tail, but lighter; the legs are of a yellowiſh brown. The female is diſtinguiſhed by a pale yellow crown; the whole plumage is leſs vivid than that of the male.

This curious little bird delights in the largeſt trees, ſuch as oaks, elms, tall pines, and firs, particularly the firſt, in which it finds both food and ſhelter; in theſe it builds its neſt, which is of a round form, having an aperture on one ſide, and is compoſed chiefly of moſs, lined with the ſofteſt down, mixed with ſlender filaments; the female lays ſix or ſeven eggs, ſcarcely larger than peas, which are white, ſprinkled with very ſmall ſpots of a dull colour. Theſe birds are very agile, and are almoſt continually in motion, fluttering from branch to branch, creeping on all ſides of the trees, clinging to them in every ſituation, and often hanging like the Titmouſe: Their food conſiſts chiefly of the ſmalleſt inſects, which they find in the crevices of the bark of trees, or catch nimbly on the wing; they alſo eat the eggs of inſects, ſmall worms, and various ſorts of ſeeds. The Golden-creſted Wren is diffuſed throughout Europe; it has alſo been met with in various parts of Aſia and America, and ſeems to bear every change of temperature, from the greateſt degree of heat to that of the ſevereſt cold: It ſtays with us the whole year; but Mr Pennant obſerves, that it croſſes annually from the Orknies to the Shetland iſles, where it breeds and returns before winter—a long flight (of ſixty miles) for ſo ſmall a bird. Its ſong is ſaid to be very melodious, but weaker than that of the common Wren; it has beſides a ſharp ſhrill cry, ſomewhat like that of the Graſhopper.

THE WREN.

KITTY WREN.

(*Motacilla troglodytes*, Lin.—*Le Troglodyte*, Buff.)

Length three inches and a half: The bill is ſlender, and a little curved; upper mandible and tips of a browniſh horn colour, the under one and edges of both dull yellow; a whitiſh line extends from the bill over the eyes, which are dark hazel; the upper parts of its plumage are of a clear brown, obſcurely marked on the back and rump with narrow double wavy lines of pale and dark brown colours; the belly, ſides, and thighs are the ſame, but more diſtinct; the throat is of a dingy white; the cheeks and breaſt the ſame, faintly dappled with brown; the quills and tail are marked with alternate bars of a reddifh brown and black; the legs are of a pale olive brown.

This diminutive little bird is very common in England, and braves our fevereft winters, which it contributes to enliven by its fprightly note. During that feafon it approaches near the dwellings of man, and takes fhelter in the roofs of houfes, barns, hay-ftacks, and holes in the walls; it continues its fong till late in the evening, and not unfrequently during a fall of fnow: In the fpring it betakes itfelf to the woods, where it builds its neft near the ground, in a low bufh, and fometimes on the turf, beneath the trunk of a tree, or in a hole in the wall; its neft is conftructed with much art, being of an oval fhape, with one fmall aperture in the fide for an entrance; it is compofed chiefly of mofs, and lined within with feathers; the female lays from ten to fixteen, and fometimes eighteen eggs, of a dirty white, dotted with red at the larger end.

THE WHITE-RUMP.

WHEATEAR.

(*Motacilla oenanthe*, Lin.—*Le Motteux, ou le cul-blanc*, Buff.)

Length five inches and a half: The bill is black; eyes hazel; from the baſe of the bill a black ſtreak extends over the eyes, cheeks, and ears, where it is pretty broad; above this there is a line of white; the top of the head, back part of the neck, and back are of a blueiſh grey; the wing coverts and quills are duſky, edged with ruſty white; the rump is perfectly white, as is alſo part of the tail; the reſt is black; the under parts are of a pale buff colour tinged with red on the breaſt; legs and feet black. In the female the white line above the eye is ſomewhat obſcure, and all the black parts of the plumage incline more to brown; neither is the rump of ſo pure a white.

This bird visits us about the middle of March, and from that time till some time in May is seen to arrive; it frequents new-tilled grounds, and never fails to follow the plough in search of insects and small worms, which are its principal food. In some parts of England great numbers are taken in snares made of horse hair, placed beneath a turf; near 2000 dozen are said to be taken annually in that way, in one district only, which are generally sold at sixpence per dozen:*—Great numbers are sent to the London markets, where they are much esteemed, being thought not inferior to the Ortolan. The White-rump breeds under shelter of a tuft or clod, in newly ploughed lands, or under stones, and sometimes in old rabbit burrows; its nest is constructed with great care; it is composed of dry grass or moss, mixed with wool, and lined with feathers; it is defended by a sort of covert, fixed to the stone or clod under which it is formed; the female generally lays five or six eggs, of a light blue, the larger end encompassed with a circle of a somewhat deeper hue. They leave us in August and September, and about that time are seen in great numbers by the sea-shore, where, probably, they subsist some little time before they take their departure. They are extended over a large portion of the globe, even as far as the southern parts of Asia.

* Pennant.

THE WHINCHAT.

(*Motacilla rubetra*, Lin.—*Le grand Traquet, ou le tarier*, Buff.)

This bird is ſomewhat larger than the Stonechat: Its bill is black; eyes hazel; the feathers on the head, neck, and back are black, edged with ruſt colour; a ſtreak of white paſſes from the bill over each eye towards the back of the head; the cheeks are blackiſh; chin white; the breaſt is of a ruſt colour; belly, vent, and thighs pale buff; each wing is croſſed by a white mark near the ſhoulder, and another ſmaller near the baſtard wing; part of the tail, at the baſe, is white, the reſt black, the two middle feathers are wholly black; the legs are black: The colours in general of the female are paler; the white ſtreak over the eye, and the ſpots on the wings, are much leſs conſpicuous; and the cheeks, inſtead of being black, partake of the colours on the head. The Whinchat is a ſolitary bird, frequenting heaths and moors; it has no ſong, but only a ſimple unvaried note, and in manners very much reſembles the Stonechat; it makes its neſt very ſimilar to that bird, and is generally ſeen in the ſame places during the ſummer months; the female lays five eggs, of a dirty white, dotted with black. In the northern parts of England it diſappears in winter; but its migration is only partial, as it is ſeen in ſome of the ſouthern counties at that ſeaſon: It feeds on worms, flies, and inſects;—about the end of ſummer it is very fat, and at that time is ſaid to be ſcarcely inferior in delicacy to the Ortolan.

THE STONECHAT.

STONE-SMITH, MOOR TITLING.

(*Motacilla rubecola*, Lin.—*Le Traquet*, Buff.)

Length near five inches: The bill is black; eyes dark hazel; the head, neck, and throat are black, faintly mixed with brown; on each ſide of the neck, immediately above the wings, there is a large white ſpot; the back and wing coverts are of a fine velvet black, margined with reddiſh brown; the quills are duſky, with pale brown edges—thoſe next the body are white at the bottom, forming a ſpot of that colour on the wings; the breaſt is of a bay colour, lighteſt on the belly; the rump white; the tail is black, the outer feathers margined with ruſt colour; the legs are black: The colours of the female are duller; the white on the ſides of the neck is not ſo conſpicuous; the breaſt and belly are much paler, and the white ſpot on the rump is wanting.

This ſolitary little bird is chiefly to be found on wild heaths and commons, where it feeds on ſmall worms and inſects of all kinds: It builds its neſt at the roots of buſhes, or underneath ſtones; it carefully conceals the entrance to it by a variety of little arts; it generally alights at ſome diſtance from it, and makes its approaches with great circumſpection, creeping along the ground in a winding direction, ſo that it is a difficult matter to diſcover its retreat; the female breeds about the end of March, and lays five or ſix eggs, of a blueiſh green, faintly ſpotted with ruſt colour. The flight of the Stonechat is low; it is almoſt continually on the wing, flying from buſh to buſh, alighting only for a few ſeconds. It remains with us the whole year, and in winter is known to frequent moiſt places, in queſt of food. Buffon compares its note to the word *wiſtrata* frequently repeated. Mr Latham obſerves, that it ſeemed to him like the clicking of two ſtones together, from whence it is probable it may have derived its name.

OF THE TITMOUSE.

This diminutive tribe is diſtinguiſhed by a peculiar degree of ſprightlineſs and vivacity, to which may be added a degree of ſtrength and courage which by no means agrees with its appearance.—Birds of this claſs are perpetually in motion; they run with great celerity along the branches of trees, ſearching for their food in every little cranny, where the eggs of inſects are depoſited, which is their favorite food: During ſpring they are frequently obſerved to be very buſy amongſt the opening buds, ſearching for the caterpillars which are lodged therein; and are thus actively employed in preventing the miſchiefs that would ariſe from a too great increaſe of theſe deſtructive inſects, whilſt, at the ſame time, they are intent on the means of their own preſervation; they will likewiſe eat ſmall pieces of raw meat, particularly fat, of which they are very fond. None of this kind have been obſerved to migrate; they ſometimes make ſhort flittings from place to place in queſt of food, but never entirely leave us.—They are very bold and daring, and will attack birds much larger than themſelves with great intrepidity. Buffon ſays, " they purſue the Owl with great fury, and that in their attacks they aim chiefly at the eyes; their actions on theſe occaſions are attended with a ſwell of the feathers, and with a ſucceſſion of violent attitudes and rapid movements, which ſtrongly mark the bitterneſs of their rage: They will ſometimes attack birds ſmaller and weaker than themſelves, which they kill, and having picked a hole in the ſkull, they eat out the brains." The neſts of moſt of this kind are conſtructed with the moſt exquiſite art, and with materials of the utmoſt delicacy; ſome ſpecies, with great ſagacity, build them at the extreme end of ſmall branches projecting over water, by which means they are effectually ſecured from the attacks of ſerpents and the ſmaller beaſts of prey. Theſe birds are very widely ſpread over every part of the old continent, from the northern parts of Europe to the Cape of Good Hope, as well as to the fartheſt parts of India, China, and Japan; they are likewiſe found throughout the vaſt continent of America, and in ſeveral of the Weſt India iſlands; They are every where prolific, even to a proverb, laying a great number of eggs, which they attend with great ſolicitude, and provide for their numerous progeny with indefatigable activity. All the Titmice are diſtinguiſhed by ſhort bills, which are conical, a little flattened at the ſides, and very ſharp pointed: The noſtrils are ſmall and round, and are generally covered by ſhort briſtly feathers, reflected from the forehead; the tongue ſeems as if cut off at the end, and terminated by ſhort filaments; the toes are divided to their origin; the back toe is very large and ſtrong.

THE GREATER TITMOUSE.

OX-EYE.

(*Parus major*, Lin.—*Le Grosse Mesange*, Buff.)

The length of this bird is about five inches: The bill is black, as are also the eyes; the head is covered with a sort of hood, of a fine deep glossy black, which extends to the middle of the neck; the cheeks are white; the belly is of a greenish yellow, divided down the middle by a line of black reaching to the vent; the back is of an olive green; rump blue grey; the quills are dusky, the greater edged with white, the lesser with pale green; the wing coverts are of a blueish ash colour; the greater coverts are tipped with white, which forms a bar across the wing; the tail is black, the exterior edge of the outer feathers is white; the legs are of a dark lead colour; claws black.

The Titmouſe begins to pair early in February; the male and female conſort for ſome time before they make their neſt, which is compoſed of the ſofteſt and moſt downy materials—they build it generally in the hole of a tree; the female lays from eight to ten eggs, which are white, ſpotted with ruſt colour. Buffon ſays, that the young brood continue blind for ſeveral days, after which their growth is very rapid, and they are able to fly in about fifteen days: After they have quitted the neſt they return no more to it, but perch on the neighbouring trees, and inceſſantly call on each other; they generally continue together till the approach of ſpring invites them to pair. We kept one of theſe birds in a cage for ſome time; it was fed chiefly with hemp-ſeed, which, inſtead of breaking with its bill, like the Linnet, it held very dexterouſly in its claws, and pecked it till it broke the outſide ſhell; it likewiſe ate raw fleſh minced ſmall, and was extremely fond of flies, which, when held to the cage, it would ſeize with great avidity: It was continually in motion during the day, and would, for hours together, dart backwards and forwards with aſtoniſhing activity. Its uſual note was ſtrong and ſimple; it had, beſides, a more varied, but very low, and not unpleaſant ſong:—During the night it reſted on the bottom of the cage.

THE BLUE TITMOUSE.

TOM TIT, BLUE-CAP, OR NUN.

(*Parus cœruleus*, Lin.—*La Mesange, bleue*, Buff.)

The length of this beautiful little bird is about four inches and a half: The bill and eyes are black; crown of the head blue, terminated behind with a line of dirty white; sides of the head white, underneath which, from the throat to the back of the neck, there is a line of dark blue; from the bill, on each side, a narrow line of black passes through the eyes; the back is of a yellowish green; coverts blue, edged with white; quills black, with pale blue edges; the tail is blue, the two middle feathers longest; the under parts of the body pale yellow; legs and claws black. The female is somewhat smaller than the male, has less blue on the head, and her colours in general are not ſo bright.

This buſy little bird is ſeen frequently in our gardens and orchards, where its operations are much dreaded by the over-anxious gardener, who fears, leſt in its purſuit after its favorite food, which is often lodged in the tender buds, that it may deſtroy them alſo, to the injury of his future harveſt —not conſidering that it is the means of deſtroying a much more dangerous enemy (the caterpillar), which it finds there: It has likewiſe a ſtrong propenſity to fleſh, and is ſaid to pick the bones of ſuch ſmall birds as it can maſter, as clean as ſkeletons. This bird is diſtinguiſhed above all the reſt of the Titmice by its rancour againſt the Owl:— The female builds her neſt in holes of walls or trees, which it lines well with feathers; ſhe lays from fourteen to twenty white eggs. If her eggs ſhould be touched by any perſon, or one of them be broken, ſhe immediately forſakes her neſt and builds again, but otherwiſe makes but one hatch in the year.

THE COLE-TITMOUSE.

(*Parus ater*, Lin.—*Le petite Charbonniere*, Buff.)

This bird is ſomewhat leſs than the laſt, and weighs only two drachms; its length is four inches: Its bill is black, as are alſo its head, throat, and part of its breaſt; from the corner of the bill, on each ſide, an irregular patch of white paſſes under the eyes, extending to the ſides of the neck; a ſpot of the ſame colour occupies the hind part of the head; the back and all the upper parts are of a greeniſh aſh colour; the wing coverts are tipped with white, which forms two bars acroſs the wing; the under parts are of a reddiſh white; legs lead colour; tail ſomewhat forked at the end.

THE MARSH TITMOUSE.

BLACK-CAPPED TITMOUSE.

(*Parus, paluſtris*, Lin.—*Le Meſange de marais*, Buff.)

Its length is ſomewhat ſhort of five inches: Its bill is black; the whole crown of the head, and part of the neck behind, are of a deep black; a broad ſtreak, of a yellowiſh white, paſſes from the beak, underneath the eye, backwards; the throat is black; the breaſt, belly, and ſides are of a dirty white; the back is aſh-coloured; quill feathers duſky, with pale edges; the tail is duſky; legs dark lead colour.

The Marſh Titmouſe is ſaid to be fond of waſps, bees, and other inſects:—It lays up a little ſtore of ſeeds againſt a ſeaſon of want: It frequents marſhy places, from whence it derives its name; its manners are ſimilar to the laſt, and it is equally as prolific.

THE LONG-TAILED TITMOUSE.

(*Parus caudatus*, Lin.—*La Meſange a longue queue*, Buff.)

The length of this bird is nearly five inches and a half, of which the tail itſelf is rather more than three inches: Its bill is very ſhort and black; eyes hazel, the orbits red; the top of the head is white, mixed with gray; through each eye there is a broad black band, which extends backwards, and unites on the hind part of the head, from whence it paſſes down the back to the rump, bordered on each ſide with dull red; the cheeks, throat, and breaſt are white; the belly, ſides, rump, and vent are of a dull roſe colour, mixed with white; the coverts of the wings are black, thoſe next the body white, edged with roſe colour; the quills are duſky, with pale edges; the tail conſiſts of feathers of very unequal lengths, the four middle feathers are wholly black, the others are white on the exterior edge; legs and claws black.

Our figure was taken from one newly ſhot, ſent us by Lieut. H. F. Gibſon. We made a drawing from a ſtuffed bird in the muſeum of the late Mr Tunſtall, at Wycliffe, in which the black band through the eyes was wholly wanting; the back of the neck was black; the back, ſides, and thighs were of a reddiſh brown, mixed with white: We ſuppoſe it may have been a female. The neſt of this bird is ſingularly curious and elegant, being of a long oval form, with a ſmall hole in the ſide as an entrance; its outſide is formed of moſs, wool, and dry graſs, curiouſly interwoven, whilſt the inſide is thickly lined with a profuſion of the ſofteſt down.*— In this comfortable little manſion the female depoſits her eggs, to the number of ſixteen or ſeventeen, which are concealed almoſt entirely among the feathers; they are about the ſize of a ſmall bean, and of a grayiſh colour, mixed with red.— This bird is not uncommon with us; it frequents the ſame places with the other ſpecies, feeds in the ſame manner, and is charged with the ſame miſdemeanor in deſtroying the buds, and probably with the ſame reaſon: It flies very ſwiftly, and from its ſlender ſhape, and the great length of its tail, it ſeems like a dart ſhooting through the air: It is almoſt conſtantly in motion, running up and down the branches of trees with great facility. The young continue with the parents, and form

* In ſome places it is called a Feather-poke.

little flocks through the winter; they utter a ſmall ſhrill cry, only as a call, but in the ſpring they are ſaid to acquire a very melodious ſong. The long-tailed Titmouſe is found in the northern regions of Europe; and, from the thickneſs of its coat, ſeems well calculated to bear the rigours of a ſevere climate. Mr Latham ſays, that it has likewiſe been brought from Jamaica, and obſerves, that it appeared as fully cloathed as in the coldeſt regions.

THE BEARDED TITMOUSE.

(*Parus biarmicus*, Lin.—*La Meſange barbue*, Buff.)

Length ſomewhat more than ſix inches: The bill is of an orange colour, but ſo delicate that it changes on the death of the bird to a dingy yellow; the eyes are alſo orange; the head and back part of the neck are of a pearl gray or light aſh colour; on each ſide of the head, from the eye, there is a black mark extending downwards on the neck, and ending in a point, not unlike a muſtachoe; the throat and fore part of the neck are of a ſilvery white; the back, rump, and tail are of a light ruſt colour, as are alſo the belly, ſides, and thighs; the breaſt is of a delicate fleſh colour; the vent black; the leſſer coverts of the wings are duſky, the greater ruſt colour, with pale edges; the quills are duſky, edged with white—thoſe next the body with ruſty on the exterior web, and with white on the inner; the baſtard wing is duſky, edged and tipped with white; the legs are black.—The female wants the black mark on each ſide of the head; the crown of the head is ruſt colour, ſpotted with black; the vent feathers are not black, but of the ſame colour with the belly.

The Bearded Titmouſe is found chiefly in the ſouthern parts of the kingdom; it frequents marſhy places, where reeds grow, on the ſeeds of which it feeds: It is ſuppoſed to breed there, though its hiſtory is imperfectly known. It is ſaid that they were firſt brought to this country from Denmark by the Counteſs of Albemarle, and that ſome of them having made their eſcape, founded a colony here; but Mr Latham, with great probability, ſuppoſes that they are ours *ab origine*, and that it is owing to their frequenting the places where reeds grow, and which are not eaſily acceſſible, that ſo little has been known of them. Mr Edwards gives a figure of this bird, and deſcribes it under the name of the Leaſt Butcher Bird.

OF THE SWALLOW.

Of all the various families of birds, which refort to this ifland for food and fhelter, there is none which has occafioned fo many conjectures refpecting its appearance and departure as the Swallow tribe:—Of this we have already hazarded our opinion in the introductory part of our work, to which we refer our readers. The habits and modes of living of this tribe are perhaps more confpicuous than thofe of any other. From the time of their arrival to that of their departure they feem continually before our eyes.—The Swallow lives habitually in the air, and performs its various functions in that element; and whether it purfues its fluttering prey, and follows the devious windings of the infects on which it feeds, or endeavours to efcape the birds of prey by the quicknefs of its motion, it defcribes lines fo mutable, fo varied, fo interwoven, and fo confufed, that they hardly can be pictured by words.—" The Swallow tribe is of all others moft inoffenfive, harmlefs, entertaining, and focial; all except one fpecies attach themfelves to our houfes, amufe us with their migrations, fongs, and marvellous agility, and clear the air of gnats and other troublefome infects, which would otherwife much annoy and incommode us. Whoever contemplates the myriads of infects that fport in the fun-beams of a fummer evening in this country, will ſoon be convinced to what a degree our atmoſphere would be choked with them, were it not for the friendly interpoſition of the Swallow tribe."* Not many attempts have been made to preſerve Swallows alive during the winter and of theſe, few have ſucceeded. The following experiments, by Mr James Pearſon of London, communicated to us by Sir John Trevelyan, Bart. are highly intereſting, and throw great light upon the natural hiſtory of the Swallow; we ſhall give them nearly in Mr Pearſon's own words.

"Five or ſix of theſe birds were taken about the latter end of Auguſt, 1784, in a bat fowling net at night; they were put ſeparately into ſmall cages, and fed with Nightingale's food: In about a week or ten days they took the food of themſelves; they were then put all together into a deep cage, four feet long, with gravel at the bottom; a broad ſhallow pan with water was placed in it, in which they ſometimes waſhed themſelves, and ſeemed much ſtrengthened by it. One day Mr Pearſon obſerved that they went into the water with unuſual eagerneſs, hurrying in and out again repeatedly, with ſuch ſwiftneſs as if they had been ſuddenly ſeized with a frenzy. Being anxious to ſee the reſult, he left them to themſelves about half an hour, and on going to the cage again, found them all huddled together in a corner of the cage, appa-

* White's Selborne.

rently dead; the cage was then placed at a proper diſtance from the fire, when two of them only recovered, and were as healthy as before—the reſt died; the two remaining ones were allowed to waſh themſelves occaſionally for a ſhort time only; but their feet ſoon after became ſwelled and inflamed, which Mr P. attributed to their perching, and they died about Chriſtmas: Thus the firſt year's experiment was in ſome meaſure loſt. Not diſcouraged by the failure of this, Mr P. determined to make a ſecond trial the ſucceeding year, from a ſtrong deſire of being convinced of the truth reſpecting their going into a ſtate of torpidity. Accordingly, the next ſeaſon, having taken ſome more birds, he put them into the cage, and in every reſpect purſued the ſame methods as with the laſt; but to guard their feet from the bad effects of the damp and cold, he covered the perches with flannel, and had the pleaſure to obſerve that the birds throve extremely well, they ſung their ſong through the winter, and ſoon after Chriſtmas began to moult, which they got through without any difficulty, and lived three or four years, regularly moulting every year at the uſual time. On the renewal of their feathers it appeared that their tails were forked exactly the ſame as in thoſe birds which return here in the ſpring, and in every reſpect their appearance was the ſame. Theſe birds, ſays Mr Pearſon, were exhibited to the Society for promoting Natural Hiſtory, on the 14th day of February, 1786, at the time they were in a deep moult, during a fevere froft, when the fnow was on the ground. Minutes of this circumftance were entered in the books of the fociety. Thefe birds died at laft from neglect during a long illnefs which Mr Pearfon had;—they died in the fummer. Mr P. concludes his very interefting account in thefe words: "Jan. 20, 1797.—I have now in my houfe, No. 21, Great Newport-ftreet, Long-Acre, four Swallows in moult, in as perfect health as any birds ever appeared to be in when moulting."

The refult of thefe experiments pretty clearly proves, that Swallows do not in any material inftance differ from other birds in their nature and propenfities; but that they leave us, like many other birds, when this country can no longer furnifh them with a fupply of their proper and natural food, and that confequently they feek it in other places, where they meet with that fupport which enables them to throw off their feathers. Swallows are found in every country of the known world, but feldom remain the whole year in the fame climate; the times of their appearance and departure in this country are well known; they are the conftant harbingers of fpring, and on their arrival all nature affumes a more chearful afpect. The bill of this genus is fhort, very broad at the bafe, and a little bent; the head is flat, and the neck fcarcely vifible; the tongue is fhort, broad, and cloven; tail moftly forked; wings long; legs fhort.

THE CHIMNEY SWALLOW.

HOUSE-SWALLOW.

(*Hirundo ruſtica*, Lin.—*L'Hirondelle domeſtique*, Buff.)

Length ſomewhat more than ſix inches: Its bill is black; eyes hazel; the forehead and chin are red, inclining to cheſtnut; the whole upper part of the body is black, reflected with a purpliſh blue on the top of the head and ſcapulars; the quills of the wings, according to their different poſitions, are ſometimes of a blueiſh black, and ſometimes of a greeniſh brown, whilſt thoſe of the tail are black, with green reflections; the fore part of the breaſt is black, and the reſt of the breaſt and belly white; the inſide and corners of the mouth are yellow; the tail is much forked, each feather, except the middle ones, is marked with an oval white ſpot on the inner web; the legs are very ſhort, delicately fine, and blackiſh.

The common Swallow makes its appearance with us ſoon after the vernal equinox, and leaves us again about the end of September: It builds its neſt generally in chimnies, in the inſide, within a few feet of the top, or under the eaves of houſes; it is curiouſly conſtructed, of a cylindrical ſhape, plaſtered with mud, mixed with ſtraw and hair, and lined with feathers; it is attached to the ſides or corners of the chimney, and is ſometimes a foot in height, open at the top; the female lays five or ſix eggs, white, ſpeckled with red. Swallows return to the ſame haunts; they build annually a new neſt, and fix it, if the place admits, above that occupied the preceding year.* We are favoured by Sir John Trevelyan, Bart. with the following curious fact:—At Camerton Hall, near Bath, a pair of Swallows built their neſt on the upper part of the frame of an old picture over the chimney, coming through a broken pane in the window of the room. They came three years ſucceſſively, and in all probability would have continued to do ſo if the room had not been put into repair, which prevented their acceſs to it. Both this bird and the Martin have generally two broods in the year, the firſt in June, the other in Auguſt, or perhaps later. We have ſeen a young Swallow, which was ſhot on the 26th

* Buffon.

of September; its length was ſcarcely five inches; its tail was ſhort, and not forked; the feathers were black, but wanted the white ſpots; its breaſt was tinged with red. Swallows frequently rooſt at night, after they begin to congregate, by the ſides of rivers and pools of water, from whence it has been ſuppoſed that they retire into that element.

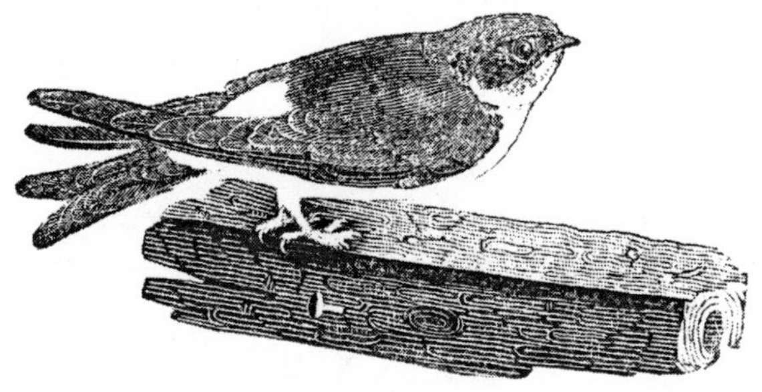

THE MARTIN.

MARTLET, MARTINET, OR WINDOW-SWALLOW.

(*Hirundo urbica*, Lin.—*L'Hirondelle à cul blanc*, Buff.)

Length about five inches and a half: The bill is black; eyes dark hazel; inſide of the mouth yellow; the top of the head, the wings, and tail are of a duſky brown; the back is black, gloſſed with blue; the rump and all the under parts of the body, from the chin to the vent, are of a pure white; the ends of the ſecondary quill feathers are finely edged with white; the legs are covered with white downy feathers down to the claws, which are white alſo, and are very ſharp and much hooked; the middle toe is much longer than the others, and is connected with the inner one as far as the firſt joint.

This bird viſits us in great numbers; it has generally two broods, ſometimes three in the year; it builds its neſt moſt frequently againſt the crags of precipices near the ſea, or by the ſides of lakes, and not unfrequently under the eaves of houſes, or cloſe by the ſides of the windows; its neſt is made of mud and ſtraw on the outſide, and is lined within with feathers; the firſt hatch the female lays five eggs, which are white, inclining to duſky at the larger end; the ſecond time ſhe lays three or four; and the third, (when that takes place) ſhe only lays two or three. During the time the young birds are confined to the neſt, the old one feeds them, adhering by the claws to the outſide; but as ſoon as they are able to fly, they receive their nouriſhment on the wing, by a motion quick and almoſt imperceptible to thoſe who are not accuſtomed to obſerve it. The Martin arrives ſomewhat later than the Swallow, and does not leave us ſo ſoon: We have obſerved them in the neighbourhood of London as late as the middle of October. Mr White, in his Natural Hiſtory of Selborne, has made ſome very judicious remarks on theſe birds, with a view to illuſtrate the time and manner of their annual migrations. The following quotation is very appoſite to that purpoſe, and ſerves to confirm the idea that the greateſt part of them quit this iſland in ſearch of warmer climates. "As the ſummer declines, the congregating flocks increaſe in numbers daily by the conſtant acceſſion of the ſecond broods; till at laſt they ſwarm in myriads upon myriads round the villages on the Thames, darkening the face of the ſky as they frequent the aits of that river, where they rooſt: They retire in vaſt flocks together about the beginning of October." He adds, "that they appeared of late years in conſiderable numbers in the neighbourhood of Selborne, for one day or two, as late as November the 3d and 6th, after they were ſuppoſed to have been gone for more than a fortnight." He concludes with this obſervation:—"Unleſs theſe birds are very ſhort-lived indeed, or unleſs they do not return to the diſtrict where they have been bred, they muſt undergo vaſt devaſtations ſomehow and ſomewhere; for the birds that return yearly bear no manner of proportion to thoſe that retire."

THE SAND MARTIN.

BANK MARTIN, OR SAND SWALLOW.

(*Hirundo riparia*, Lin.—*L'Hirondelle de rivage*, Buff.)

LENGTH about four inches and three quarters: The bill is of a dark horn colour; the head, neck, breaſt, and back are of a mouſe colour; over each eye there is a light ſtreak; the throat and fore part of the neck are white, as are alſo the belly and vent; the wings and tail are brown; the legs are dark brown, and are furniſhed with feathers behind, which reach as far as the toes.

This is the ſmalleſt of all our Swallows, as well as the leaſt numerous of them: It frequents the ſteep ſandy banks in the neighbourhood of rivers, in the ſides of which it makes deep holes, and places the neſt at the end; it is careleſsly conſtructed of ſtraw, dry graſs, and feathers; the female lays five or ſix white eggs, almoſt tranſparent, and is ſaid to have only one brood in the year.

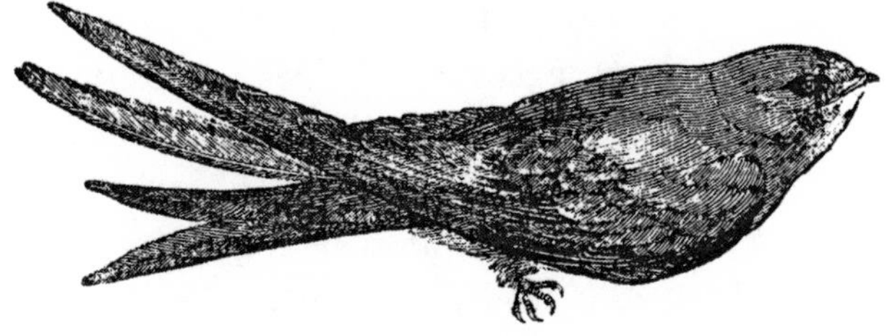

THE SWIFT.

BLACK MARTIN, OR DEVILING.

(*Hirundo apus*, Lin.—*Le Martinet noir*, Buff.)

Length near eight inches: Bill black; eyes hazel; its general colour is that of a ſooty black, with greeniſh reflections; the throat is white; the wings are long, meaſuring, from tip to tip, about eighteen inches; the tail is much forked; the legs are of a dark brown colour, and very ſhort; the toes ſtand two and two on each ſide of the foot, and conſiſt of two phalanges or joints only, which is a conformation peculiar to this bird. The female is rather leſs than the male, her plumage inclines more to brown, and the white on the throat is leſs diſtinct.

The Swift arrives later and departs ſooner than any of the tribe, from whence it is probable that it has a longer journey to take than the others; it is larger, ſtronger, and its flight is more rapid than any of its kindred tribes, and it has but one brood in the year, ſo that the young ones have time to gain ſtrength enough to accompany the parent birds in their diſtant excurſions: They have been noticed at the Cape of Good Hope, and probably viſit the more remote regions of Aſia. Swifts are almoſt continually on the wing; they fly higher, and wheel with bolder wing than the Swallows, with which they never intermingle. The life of the Swift ſeems to be divided into two extremes; the one of the moſt violent exertion, the other of perfect inaction; they muſt either ſhoot through the air, or remain cloſe in their holes; they are ſeldom ſeen to alight; but, if by any accident they ſhould fall upon a piece of even ground, it is with difficulty they can recover themſelves, owing to the ſhortneſs of their feet, and the great length of their wings. They are ſaid to avoid heat, and for this reaſon paſs the middle of the day in their holes; in the morning and evening they go out in queſt of proviſion; they then are ſeen in large flocks, deſcribing an endleſs ſeries of circles upon circles, ſometimes in cloſe ranks, purſuing the direction of a ſtreet, and ſometimes whirling round a large edifice, all ſcreaming together; they often glide along without ſtirring their wings, and on a ſudden they move them with frequent and quickly repeated ſtrokes. Swifts build their neſts in elevated places; lofty ſteeples and high towers are generally preferred; ſometimes they build under the arches of bridges, where, though the elevation is not great, it is difficult of acceſs; the neſt is composed of a variety of materials, ſuch as dry graſs, moſs, hemp, bits of cord, threads of ſilk and linen, ſmall ſhreds of gauze, of muſlin, feathers, and other light ſubſtances which they chance to find in the ſweepings of towns.* It is difficult to conceive how theſe birds, which are never ſeen to alight on the ground, gather theſe materials; ſome have ſuppoſed that they catch them in the air as they are carried up by the wind; others, that they raiſe them by glancing along the ſurface of the ground; whilſt others aſſert, with more probability, that they often rob the Sparrow of its little hoard, and frequently occupy the ſame hole after driving out the former poſſeſſor: The female lays five white eggs, rather pointed and ſpindle-ſhaped; the young ones are hatched about the latter end of May; they begin to fly about the middle of June, and ſhortly after abandon their neſts—after which the parents ſeem no more to regard them.—Swifts begin to aſſemble, previous to their departure, early in July; their numbers daily increaſe, and large bodies of them appear together; they ſoar higher in the air, with ſhriller cries, and fly differently from their uſual mode;—theſe meetings continue till towards the middle of Auguſt, after which they are no more ſeen.

* Buffon.

THE NIGHT-JAR.

GOAT-SUCKER, DORR-HAWK, OR FERN OWL.

(*Caprimulgus Europeus*, Lin.—*L'Engoulivent*, Buff.)

The length of this bird is about ten inches and a half: The bill is ſmall, flat, and ſomewhat hooked at the tip, and is furniſhed on each ſide of the upper mandible with ſeveral ſtrong briſtles, whereby it ſecures its prey; the lower jaw is edged with a white ſtripe, which extends backward towards the head; the eyes are large, full, and black; the plumage is beautifully freckled and powdered with browns of various hues, mixed with ruſt colour and white, but ſo diverſified as to exceed all deſcription. The male is diſtinguiſhed by an oval ſpot of white on the inner webs of the three firſt quill feathers, and at the ends of the two outermoſt feathers of the tail; the legs are ſhort, rough, and ſcaly, and feathered below the knee; the toes are connected by a membrane as far as the firſt joint; the middle one is conſiderably larger than the reſt, and the claw is ſerrated in one ſide.

To avoid as much as poſſible perpetuating error, we have dropped the term Goat-ſucker, which has no foundation but in ignorance and ſuperſtition, and have adopted one, which, though not univerſally known, bears ſome analogy to the nature and qualities of the bird to which it relates, both with reſpect to the time of its appearance, which is always in the duſk of the evening, in ſearch of its prey, as well as to the jarring noiſe which it utters whilſt at reſt perching on a tree, and by which it is peculiarly diſtinguiſhed. The Night-jar is found in every part of the old continent, from Siberia to Greece, Africa, and India; it arrives in this country about the latter end of May, being one of our lateſt birds of paſſage, and departs ſome time in the latter end of Auguſt or the beginning of September; it is no where numerous, and never appears in flocks: Like the Owl, it is ſeldom ſeen in the day-time, unleſs diſturbed, or in dark and gloomy days, when its eyes are not dazzled by the bright rays of the ſun: It feeds on inſects, which it catches on the wing; it is a great deſtroyer of the cock-chafer or dor-beetle, from whence in ſome places it is called the Dor-hawk: Six of theſe inſects have been found in its ſtomach, beſides four or five large-bodied moths. Mr White ſuppoſes that its foot is uſeful in taking its prey, as he obſerved that it frequently put forth its leg whilſt on the wing, with which it ſeemed to convey ſomething to its mouth. Theſe birds frequent moors and wild heathy tracts abounding with ferns; they make no neſt, but the female depoſits her eggs on the ground; ſhe lays only two or three, which are of a dull white, ſpotted with brown. Birds of this kind are ſeen moſt frequently towards autumn; their motions are irregular and rapid, ſometimes wheeling in quick ſucceſſion round a tree or other object, diving at intervals as if to catch their prey, and then riſing again as ſuddenly. When perched, the Night-jar ſits uſually on a bare twig, its head lower than its tail, and in this attitude utters its jarring note; it is likewiſe diſtinguiſhed by a ſort of buzzing which it makes while on the wing, which has been compared to the noiſe cauſed by the quick rotation of a ſpinning-wheel, from which, in ſome places, it is called the Wheel-bird; ſometimes it utters a ſmall plaintive note or ſqueak, which it repeats four or five times in ſucceſſion; the latter, probably, is its note of call by which it invites the female, and which it has been obſerved to utter when in purſuit of her. Buffon ſays, that it does not perch like other birds, ſitting acroſs the branch, but lengthwiſe. It is a ſolitary bird, and is generally ſeen alone, two being ſeldom found together, but ſitting at a little diſtance from each other.

OF THE DOVE KIND.

The various families which conſtitute this beautiful kind are diſtinguiſhed by ſhades and gradations ſo minute as to exceed all deſcription. Of theſe by far the largeſt portion are the willing attendants on man, and dependent on his bounty; but when we conſider the lightneſs of their bodies, the great ſtrength of their wings, and the amazing rapidity of their flight, it is a matter of wonder that they ſhould ſubmit even to a partial kind of domeſtication, or occupy thoſe tenements fitted up for the purpoſe of breeding and rearing their young. It muſt be obſerved, however, that in theſe they live rather as voluntary captives, or tranſient gueſts, than permanent or ſettled Inhabitants, enjoying a conſiderable portion of that liberty they ſo much delight in: On the ſlighteſt diſappointment they abandon their manſion with all its conveniences; ſome take refuge in the woods, where, impelled by inſtinct, they reſume their native manners; others ſeek a ſolitary lodgment in the holes of old walls, or unfrequented towers; whilſt others, ſtill more domeſticated, ſeldom leave their dwelling, and only roam abroad to ſeek amuſement, or to procure ſubſiſtence.

Of theſe the varieties and intermixtures are innumerable, and partake of all thoſe varied hues which are the conſtant reſult of domeſtication.—

The manners of pigeons are well known, few ſpecies being more univerſally diffuſed; and having a very powerful wing, they are enabled to perform very diſtant journies; accordingly wild and tame pigeons occur in every climate, and although they thrive beſt in warm countries, yet with care they ſucceed alſo in very northern latitudes: Every where their manners are gentle and lively; they are fond of ſociety, and the very emblem of connubial attachment; they are faithful to their mates, whom they ſolicit with the ſofteſt cooings, the tendereſt careſſes, and the moſt graceful movements. The exterior form of the Pigeon is beautiful and elegant; the bill is weak, ſtraight, and ſlender, and has a ſoft protuberance at the baſe, in which the noſtrils are placed; the legs are ſhort and red, and the toes divided to the origin.

THE WILD PIGEON.

STOCK DOVE.

(*Columba œnas*, Lin.—*Le Biſet*, Buff.)

Length fourteen inches: Bill pale red; the head, neck, and upper part of the back are of a deep blue gray colour, reflected on the ſides of the neck with gloſſy green and gold; the breaſt is of a pale reddiſh purple, or vinous colour; the lower part of the back and rump light gray or aſh colour, as are alſo the belly, thighs, and under tail coverts; the primary quill feathers are duſky, edged with white, the others gray, marked with two black ſpots on the exterior webs, forming two bars acroſs each wing; the tail is aſh colour and black at the end, the lower half of the two outermoſt feathers is white; the legs are red; claws black. —The Stock Dove, Rock Pigeon, and Wood Pigeon, with ſome ſmall differences, may be included under the ſame denomination, and are probably the origin of moſt of thoſe beautiful varieties which, in a ſtate of domeſtication, are dependent upon man for food.

Wild Pigeons are ſaid to migrate in large flocks into England, at the approach of winter, from the northern regions, and return in the ſpring; many of them, however, remain in this country, only changing their quarters for the purpoſe of procuring their food: They build their neſts in the hollows of decayed trees, and commonly have two broods in the year. In a ſtate of domeſtication their increaſe is prodigious; and though they never lay more than two eggs at a time, yet, allowing them to breed nine times in the year, the produce of a ſingle pair, at the expiration of four years, may amount to the enormous number of 14,762.* The male and female perform the office of incubation by turns, and feed their young by caſting up the proviſions out of their ſtomachs into the mouths of the young ones. To deſcribe the numerous varieties of the domeſtic Pigeon would exceed the limits of our work; we ſhall therefore barely mention the names of the moſt

* Stillingfleet's Tracts.

noted amongſt them, ſuch as Tumblers, Carriers, Jacobines, Croppers, Powters, Runts, Turbits, Shakers, Smiters, Owls, Nuns, &c. Of theſe the Carrier Pigeon is the moſt remarkably deſerving of notice, having been made uſe of in very early times as the means of conveying intelligence on the moſt trying and important occaſions, and with an expedition and certainty which could be equalled by no other. The Pigeon uſed on theſe occaſions was taken from the place to which the advices were to be communicated, and the letters being tied under its wings, the bird was let looſe, and in ſpite of ſurrounding armies and every obſtacle that would have effectually prevented any other means of conveyance, guided by inſtinct alone, it returned directly home, where the intelligence was ſo much wanted. There are various inſtances on record of theſe birds having been employed during a ſiege to convey an account of its progreſs, of the ſituation of the beſieged, and of the probable means of relief: Sometimes they were the peaceful bearer of glad tidings to the anxious lover, and to the merchant of the no leſs welcome news of the ſafe arrival of his veſſel at the deſired port.

THE RING DOVE.

CUSHAT, OR QUEEST.

(*Columba palumbus*, Lin.—*Le Pigeon ramier*, Buff.)

This is the largeſt of all the Pigeon tribe, and meaſures above ſeventeen inches in length: The bill is of a pale red colour, the noſtrils being covered with a mealy red fleſhy membrane; the eyes are pale yellow; the upper parts of the body are of a blueiſh aſh colour, deepeſt on the upper part of the back, the lower part of which, the rump, and fore part of the neck and head, are of a pale aſh colour; the lower part of the neck and breaſt are of a vinous aſh colour; and the belly, thighs, and vent are of a dull white; on the hind part of the neck is a ſemicircular line of white—from whence its name—above and beneath which, the feathers are gloſſy, and of a changeable hue in different lights; the greater quills are duſky, and all of them, except the outermoſt, edged with white; from the point of the wing a white line extends downwards, paſſing above the baſtard wing; the tail is aſh colour, tipped with black; the legs are red, and partly covered with feathers; the claws are black.—Our figure was taken from ſpecimens ſent us by John Trevelyan, Eſq. and Mr Bailey of Chillingham.

The Ring Dove is very generally diffuſed throughout Europe: It is ſaid to be migratory, but that it does not leave us entirely we are well convinced, as we have frequently ſeen them during the winter on the banks of the Tyne, where they conſtantly breed: The neſt is compoſed of ſmall twigs, ſo looſely put together, that the eggs may be ſeen through it from below. The female lays two eggs, and is generally ſuppoſed to have two broods in the year: They feed on wild fruits, herbs, and grain of all kinds; they will likewiſe eat turnips, which give their fleſh an unpleaſant flavour. The Ring Dove has a louder and more plaintive ſort of cooing than the common Pigeon, but is not heard except in pairing time, or during fine weather; when it rains, or in the gloom of winter, it is generally ſilent. Their fleſh is excellent, eſpecially when young.

THE TURTLE DOVE.

(*Columba turtur*, Lin.—*La Tourterelle*, Buff.)

Length ſomewhat more than twelve inches: The bill is brown; eyes yellow, encompaſſed with a crimſon circle; the top of the head is aſh colour, mixed with olive; each ſide of the neck is marked with a ſpot of black feathers, tipped with white; the back is aſh colour, each feather margined with reddiſh brown; wing coverts and ſcapulars reddiſh brown, ſpotted with black; quill feathers duſky, with pale edges; the fore part of the neck and breaſt are of a light purpliſh red; the belly, thighs, and vent white; the two middle feathers of the tail are brown, the others duſky, tipped with white, the two outermoſt edged and tipped with white; the legs are red.—One of theſe birds, which was ſent us by the Rev. Henry Ridley, was ſhot out of a flock at Preſtwick-Carr, in Northumberland, in the month of September, 1794: It agreed in every reſpect with the common Turtle, excepting the mark on each ſide of the neck, which was wholly wanting. We ſuppoſe it to have been a young bird. The note of the Turtle Dove is ſingularly tender and plaintive: In addreſſing his mate the male makes uſe of a variety of winning attitudes, cooing at the ſame time in the moſt gentle and ſoothing accents; on this account the Turtle Dove has been repreſented, in all ages, as the moſt perfect emblem of connubial attachment and conſtancy. The Turtle arrives late in the ſpring, and departs about the latter end of Auguſt: It frequents the thickeſt and moſt ſheltered parts of the woods, where it builds its neſt on the higheſt trees: The female lays two eggs, and has only one brood in this country, but in warmer climates it is ſuppoſed to breed ſeveral times in the year. Turtles are pretty common in Kent, where they are ſometimes ſeen in flocks of twenty or more, frequenting the pea fields, and are ſaid to do much damage. Their ſtay with us ſoldom exceeds more than four or five months, during which time they pair, build their neſts, breed and rear their young, which are ſtrong enough to join them in their retreat.

OF THE GALLINACEOUS KIND.

We are now to ſpeak of a very numerous and uſeful claſs of birds, which, by the bountiful diſpoſition of Providence, is diffuſed throughout every country of the world, affording every where a plentiful and grateful ſupply of the moſt delicate, wholeſome, and nutritious food. A large portion of theſe ſeem to have left their native woods to crowd around the dwellings of man, where, ſubſervient to his purpoſe, they ſubſiſt upon the pickings of the barn-yard, the ſtable, or the dunghill; a chearful, active race, which enlivens and adorns the rural ſcene, and requires no other care than the foſtering hand of the houſewife to ſhelter and protect it. Some kinds, ſuch as the Partridge, the Pheaſant, and the like, are found only in cultivated places, at no great diſtance from the habitations of men: and, although they have not ſubmitted to his dominion, they are nevertheleſs ſubject to his controuling power, and the object of his keeneſt attention and purſuit:—Whilſt others, taking a wider range, find food and ſhelter in the deepeſt receſſes of the woods and foreſts, ſometimes ſubſiſting upon wild and heathy mountains, or among rocks and precipices of the moſt difficult acceſs. The characters of this claſs are generally well known; they are diſtinguiſhed above all others for the whiteneſs of their fleſh;

their bodies are large and bulky, and their heads comparatively ſmall; the bill in all of them is ſhort, ſtrong, and ſomewhat curved; their wings are ſhort and concave, and ſcarcely able to ſupport their bodies, on which account they ſeldom make long excurſions; their legs are ſtrong, and are furniſhed with a ſpur or knob behind.—Birds of this kind are extremely prolific, and lay a great number of eggs; the young follow the mother as ſoon as hatched, and immediately learn to pick up the food which ſhe is moſt aſſiduous in ſhewing them; on this account ſhe generally makes her neſt on the ground, or in places of eaſy acceſs to her young brood.

Our gallant Chanticleer holds a diſtinguiſhed rank in this claſs of birds, and ſtands foremoſt in the liſt of our domeſtic tribes; on which account we ſhall place him at the head.

THE DOMESTIC COCK.

(*Phaſianus Gallus*, Lin.—*Le Coq*, Buff.)

The Cock, like the Dog, in his preſent ſtate of domeſtication differs ſo widely from his wild original, as to render it a difficult matter to trace him back to his primitive ſtock; however it is generally agreed that he is to be found in a ſtate of nature in the foreſts of India, and in moſt of the iſlands of the Indian ſeas. The varieties of this ſpecies are endlefs, every country, and almoft every diftrict of each country producing a different kind. From Afia, where they are fuppofed to have originated, they have been diffufed over every part of the inhabited world. America was the laft to receive them. It has been faid that they were firft introduced into Brazil by the Spaniards; they are now as common in all the inhabited parts of that vaft continent as with us. Of thofe which have been felected for domeftic purpofes in this country, the principal are,

1. The Crefted Cock, of which there are feveral varieties, fuch as the white-crefted black ones; the black-crefted white ones; the gold and filver ones, &c.

2. The Hamburgh Cock, named alfo Velvet Breeches, becaufe its thighs and belly are of a foft black.* This is a very large kind, and much ufed for the table.

3. The Bantam, or Dwarf Cock, a diminutive but very fpirited breed: Its legs are furnifhed with long feathers, which reach to the ground behind; it is very courageous, and will fight with one much ftronger than itfelf.

4. The Frizzled Cock. The feathers in this are fo curled up that they feem reverfed, and to ftand in oppofite directions: They are originally

* Buffon.

from the ſouthern parts of Aſia, and when young are extremely ſenſible of cold: They have a diſordered and unpleaſant appearance, but are in much eſteem for the table.

We ſhall finiſh our liſt with the Engliſh Game-Cock, which ſtands unrivalled by thoſe of any other nation for its invincible courage, and on that account is made uſe of as the inſtrument of the cruel ſport of cock-fighting. To trace this cuſtom to its origin we muſt look back into barbarous times, and lament that it ſtill continues the diſgrace of an enlightened and philoſophic age. The Athenians allotted one day in the year to cock-fighting; the Romans are ſaid to have learned it from them; and by that warlike people it was firſt introduced into this iſland. Henry VIII. was ſo attached to the ſport, that he cauſed a commodious houſe to be erected for that purpoſe, which, though it is now applied to a very different uſe, ſtill retains the name of the Cock-pit. The Chineſe and many of the nations of India are ſo extravagantly fond of this unmanly ſport, that, during the paroxyſms of their phrenzy, they will ſometimes riſk not only the whole of their property, but their wives and children on the iſſue of a battle.

The appearance of the Game-cock, when in his full plumage, and not mutilated for the purpoſe of fighting, is ſtrikingly beautiful and animated: His head, which is ſmall, is adorned with a beautiful red comb and wattles; his eyes ſparkle with fire and his whole demeanour befpeaks boldnefs and freedom; the feathers on his neck are long, and fall gracefully down upon his body, which is thick, firm, and compact; his tail is long, and forms a beautiful arch behind, which gives a grace to all his motions; his legs are ftrong, and are armed with fharp fpurs, with which he defends himfelf and attacks his adverfary. When furrounded by his females, his whole afpect is full of animation; he allows of no competitor, but on the approach of a rival he rufhes forward to inftant combat, and either drives him from the field, or perifhes in the attempt. The Cock is very attentive to his females, hardly ever lofing fight of them; he leads, defends, and cherifhes them, collects them together when they ftraggle, and feems to eat unwillingly till he fees them feeding around him; when he lofes them he utters his griefs, and from the different inflexions of his voice, and the various fignificant geftures which he makes, one would be led to conclude that it is a fpecies of language which ferves to communicate his fentiments. The fecundity of the hen is great; fhe lays generally two eggs in three days, and continues to lay through the greateft part of the year, except during the time of moulting, which lafts about two months. After having laid about twenty-five or thirty eggs, fhe prepares for the painful tafk of incubation, and gives the moft certain indications of her wants by her cries and the violence of her emotions. If ſhe ſhould be deprived of her own eggs, which is frequently the caſe, ſhe will cover thoſe of any other kind, or even fictitious ones of ſtone or chalk, by which means ſhe waſtes herſelf in fruitleſs efforts. A ſitting hen is a lively emblem of the moſt affectionate ſolicitude and attention; ſhe covers her eggs with her wings, foſters them with a genial warmth, changing them gently, that all parts may be properly heated; ſhe ſeems to perceive the importance of her employment, and is ſo intent in her occupation, that ſhe neglects, in ſome meaſure, the neceſſary ſupplies of food and drink; ſhe omits no care, overlooks no precaution, to complete the exiſtence of the little incipient beings, and to guard againſt the dangers that threaten them. Buffon, with his uſual elegance, obſerves, "that the condition of a ſitting hen, however inſipid it may appear to us, is perhaps not a tedious ſituation, but a ſtate of continual joy; ſo much has Nature connected raptures with whatever relates to the multiplication of her creatures!"

For a curious account of the progreſs of incubation, in the developement of the chick, we refer our readers to the above-mentioned author, who has given a minute detail of the ſeveral appearances which take place, at different ſtated periods, till the young chick is ready to break the ſhell and come forth. In former times the Egyptians, and in later days philoſophers, have ſucceeded in hatching eggs without the aſſiſtance of the hen, and that in great numbers at once, by means of artificial heat, correſponding with the warmth of the hen: The eggs are placed in ovens, to which an equal and moderate degree of heat is applied, and every kind of moiſture or pernicious exhalation carefully avoided—by which means, and by turning the eggs ſo that every part may enjoy alike the requiſite heat, hundreds may be produced at the ſame time.

THE PHEASANT.

(*Phaſianus Colchicus*, Lin.—*Le Faiſan*, Buff.)

Is of the ſize of the common Cock: The bill is of a pale horn colour; the noſtrils are hid under an arched covering; the eyes are yellow, and are ſurrounded by a naked warty ſkin, of a beautiful ſcarlet, finely ſpotted with black; immediately under each eye there is a ſmall patch of ſhort feathers, of a dark gloſſy purple; the upper parts of the head and neck are of a deep purple, varying to gloſſy green and blue; the lower parts of the neck and breaſt are of a reddiſh cheſtnut, with black indented edges; the ſides and lower part of the breaſt are of the ſame colour, with pretty large tips of black to each feather, which in different lights vary to a gloſſy purple; the belly and vent are duſky; the back and ſcapulars are beautifully variegated with black and white, or cream colour ſpeckled with black, and mixed with deep orange, all the feathers being edged with black; on the lower part of the back there is a mixture of green; the quills are duſky, freckled with white; wing coverts brown, gloſſed with green, and edged with white; rump plain reddiſh brown; the two middle feathers of the tail are about twenty inches long, the ſhorteſt on each ſide leſs than five, of a reddiſh brown colour, marked with tranſverſe bars of black; the legs are duſky, with a ſhort blunt ſpur on each; between the toes there is a ſtrong membrane.

The female is leſs, and does not exhibit that variety and brilliancy of colours which diſtinguiſh the male: The general colours are light and dark brown, mixed with black; the breaſt and belly finely freckled with ſmall black ſpots on a light ground; the tail is ſhort, and barred ſomewhat like the male; the ſpace round the eye is covered with feathers.

The Ring Pheaſant is a fine variety of this breed; its only difference conſiſts in a white ring, which encircles the lower part of the neck; the colours of the plumage in general are likewiſe more diſtinct and vivid. A fine ſpecimen of this bird was ſent us by the Rev. Wm Turner, of Newcaſtle, from which our figure was engraven. They are ſometimes met with in the neighbourhood of Alnwick, whither they were brought by his Grace the Duke of Northumberland. That they intermix with the common breed is very obvious, as in thoſe we have ſeen the ring has been more or leſs diſtinct; in ſome hardly viſible, and in others a few feathers only, marked with white, appear on each ſide of the neck, forming a white ſpot. It is much to be regretted that this beautiful breed is likely ſoon to be deſtroyed by thoſe who purſue every ſpecies of game with an avaricious and indiſcriminating rapacity.

There are great varieties of Pheaſants, of extraordinary beauty and brilliancy of colours; many of theſe, brought from the rich provinces of China, are kept in aviaries in this kingdom; the Common Pheaſant is likewiſe a native of the eaſt, and is the only one of its kind that has multiplied in our iſland. Pheaſants are generally found in low woody places, on the borders of plains, where they delight to ſport; during the night they perch on the branches of trees: They are very ſhy birds, and do not aſſociate together, except during the months of March and April, when the male ſeeks the female; they are then eaſily diſcoverable by the noiſe which they make in crowing and clapping their wings, which may be heard at ſome diſtance. The hen breeds on the ground like the Partridge, and lays from twelve to fifteen eggs, which are ſmaller than thoſe of the Common Hen; the young follow the mother as ſoon as ever they are freed from the ſhell. During the breeding ſeaſon the cocks will ſometimes intermix with the Common Hen, and produce a hybrid breed, of which we have known ſeveral inſtances.

THE TURKEY.

(*Meleagris Gallopavo*, Lin.—*Le Dindon*, Buff.)

It ſeems to be generally allowed that this bird was originally brought from America, and in its wild ſtate is conſiderably larger than our domeſtic Turkies: Its general colour is black, variegated with bronze and bright gloſſy green, in ſome parts changing to purple; the quills are green gold, black towards the ends, and tipped with white; the tail conſiſts of eighteen feathers, of a brown colour, mottled and tipped with black; the tail coverts are waved with black and white; on the breaſt there is a tuft of black hairs, eight inches in length: In other reſpects it reſembles the domeſtic Turkey, in having a bare red carunculated head and neck, a fleſhy dilatable appendage hanging over the bill, and a ſhort blunt ſpur or knob at the back part of the leg.

Tame Turkies, like every other animal in a ſtate of domeſtication, are of various colours; of theſe the prevailing one is dark grey, inclining to black, with a little white towards the ends of the feathers: ſome are perfectly white; others black and white; there is alſo a beautiful variety of a fine deep copper colour, with the greater quills pure white; the tail of a dirty white: In all of them the tuft of black hair on the breaſt is prevalent. Turkies are bred in great numbers in Norfolk, Suffolk, and other counties, from whence they are driven to the London markets in flocks of ſeveral hundreds each. The drivers manage them with great facility, by means of a bit of red rag tied to a long pole, which, from the antipathy theſe birds bear to that colour, acts as a ſcourge, and effectually anſwers the purpoſe. The motions of the Turkey, when agitated with deſire or inflamed with rage, are very ſimilar to thoſe of the Peacock; it erects its train, and ſpreads it like a fan, whilſt its wings droop and trail on the ground, uttering at the ſame time a dull hollow ſound; it ſtruts round and round with ſolemn pace, aſſumes all the dignity of the moſt majeſtic of birds, and thus expreſſes its attachment to its females, or its resentment to those objects which have excited its indignation. The Hen Turkey begins to lay early in the spring; she is very attentive to the business of incubation, and will produce fifteen or sixteen chicks at one time, but seldom has more than one hatch in a season in this climate. Young Turkies, after their extrication from the shell, are very tender, and require great attention in rearing them; they are subject to a variety of diseases from cold, rain, and dews; even the sun itself, when they are exposed to its more powerful rays, is said to occasion almost immediate death. As soon as they are sufficiently strong, they are abandoned by the mother, and are then capable of enduring the utmost rigour of our winters.

THE PEACOCK.

(*Pavo criſtatus* Lin.—*Le Paon,* Buff.)

To deſcribe the inimitable beauties of this elegant bird in adequate terms, would be a taſk of no ſmall difficulty. "Its matchleſs plumage," ſays Buffon, "ſeems to combine all that delights the eye in the ſoft and delicate tints of the fineſt flowers; all that dazzles it in the ſparkling luſtre of the gems; and all that aſtoniſhes it in the grand diſplay of the rainbow." Its head is adorned with a tuft, conſiſting of twenty-four feathers, whoſe ſlender ſhafts are furniſhed with webs only at the ends, painted with the moſt exquiſite green, mixed with gold; the head, throat, neck, and breaſt, are of a deep blue, gloſſed with green and gold; the back the ſame, tinged with bronze; the ſcapulars and leſſer wing coverts are of a reddiſh cream colour, variegated with black; the middle coverts deep blue, gloſſed with green and gold; the greater coverts and baſtard wing are of a reddiſh brown, as are alſo the quills, ſome of which are variegated with black and green; the belly and vent are black, with a greeniſh hue: But the diſtinguiſhing character of this ſingular bird is its train, which riſes juſt above the tail, and, when erected, forms a fan of the moſt reſplendent hues; the two middle feathers are ſometimes four feet and a half long, the others gradually diminiſhing on each ſide; the ſhafts, which are white, are furniſhed from their origin nearly to the end with parted filaments of varying colours, ending in a flat vane, which is decorated with what is called *the eye*. "This is a brilliant ſpot, enamelled with the moſt enchanting colours; yellow, gilded with various ſhades;

green, running into blue and bright violet, varying according to its different poſitions; the whole receiving additional luſtre from the colour of the centre, which is a fine velvet black." When pleaſed or delighted, and in ſight of his females, the Peacock erects his tail, and diſplays all the majeſty of its beauty; all his movements are full of dignity; his head and neck bend nobly back his pace is ſlow and ſolemn, and he frequently turns ſlowly and gracefully round, as if to catch the ſun-beams in every direction, and produce new colours of inconceiveable richneſs and beauty, accompanied at the ſame time with a hollow murmuring voice expreſſive of deſire.

The Peahen is ſomewhat leſs than the cock, and though furniſhed both with a train and creſt, they are deſtitute of thoſe dazzling beauties which diſtinguiſh the male: She lays five or ſix eggs, of a whitiſh colour: For this purpoſe ſhe chuſes ſome ſecret ſpot where ſhe can conceal them from the male, who is apt to break them; ſhe ſits from twenty-five to thirty days, according to the temperature of the climate, and the warmth of the ſeaſon. Peacocks have been originally brought from the diſtant provinces of India, and from thence have been diffuſed over every part of the world. —The firſt notice that has been taken of them is to be found in holy writ,* where we are told,

* 2d Chron. ix. 21.

they made part of the cargoes of the rich and valuable fleet which every three years imported the treaſures of the Eaſt to Solomon's court. They are ſometimes found in a wild ſtate in many parts of Aſia and Africa: The largeſt and fineſt are ſaid to be met with in the neighbourhood of the Ganges, and on the fertile plains of India, where they grow to a great ſize; under the influence of that luxuriant climate this beautiful bird exhibits its dazzling colours, which ſeem to vie with the gems and precious ſtones produced in thoſe delightful regions. In colder climates they require great care in rearing, and do not obtain their full plumage till the third year. In former times they were conſidered as a delicacy, and made a part of the luxurious entertainment of the Roman voluptuaries.

White Peacocks are not uncommon in England, the eyes of the train not excepted, which are barely viſible, and may be traced by a different undulation of ſhade upon the pure white of the tail. It is a very ſingular circumſtance, that the females of this ſpecies have been ſometimes known to aſſume the appearance of the male, by a total change of colour. This is ſaid to take place after ſhe has done laying. A bird of this kind is preſerved in the Leverian Muſeum.

THE PINTADO.

GUINEA HEN, OR PEARLED HEN.

(*Numidia Meleagris*, Lin.—*La Pintade*, Buff.)

This bird is ſomewhat larger than the common Hen: Its head is bare of feathers, and covered with a naked ſkin, of a blueiſh colour; on the top is a callous protuberance, of a conical form; at the baſe of the upper bill, on each ſide, there hangs a looſe wattle, which in the female is red, and in the male of a blueiſh colour; the upper part of the neck is almoſt naked, being very thinly furniſhed with a few ſtraggling hairy feathers; the ſkin is of a light aſh colour; the lower part of the neck is covered with feathers of a purple hue; the general colour of the plumage is a dark blueiſh grey, ſprinkled with round white ſpots of different ſizes, reſembling pearls—hence it has been called the Pearled Hen; its wings are ſhort, and its tail pendulous, like that of the Partridge; its legs are of a dark colour.

This ſpecies, which is now very common in this country, was originally brought from Africa, from whence it has been diffuſed over every part of Europe, the Weſt Indies, and America: It formed a part of the Roman banquets, and is now much eſteemed as a delicacy, eſpecially the young birds. The female lays a great number of eggs, which ſhe frequently ſecretes till ſhe has produced her young brood: The egg is ſmaller than that of a common Hen, and of a rounder ſhape; it is very delicious eating. The Pintado is a reſtleſs and very clamorous bird; it has a harſh, creaking note, which is very grating and unpleaſant; it ſcrapes the ground like the Hen, and rolls in the duſt to free itſelf from inſects; during the night it perches on high places; if diſturbed, it alarms every thing within hearing by its unceaſing cry. In its natural ſtate of freedom it is ſaid to prefer marſhy places.

THE WOOD GROUSE.

COCK OF THE WOOD, OR CAPERCAILE.

(*Tetrao urogallus*, Lin.—*Le grand Coq de Bruyere*, Buff.)

This bird is as large as a Turkey, is about two feet nine inches in length, and weighs from twelve to fifteen pounds: The bill is very ſtrong, convex, and of a horn colour; over each eye there is a naked ſkin, of a bright red colour; the eyes are hazel; the noſtrils are ſmall, and almoſt hid under a covering of ſhort feathers, which extend under the throat, and are there much longer than the reſt, and of a black colour; the head and neck are elegantly marked with ſmall tranſverſe lines of black and grey, as are alſo the back and wings, but more irregularly; the breaſt is black, richly gloſſed with green on the upper part, and mixed with a few white feathers on the belly and thighs; the ſides are marked like the neck; the tail conſiſts of eighteen feathers, which are black, thoſe on the ſides being marked with a few white ſpots; the legs are very ſtout, and covered with brown feathers; the toes are furniſhed on each ſide with a ſtrong pectinated membrane. The female is conſiderably leſs than the male, and differs greatly in her colours: The throat is red; the tranſverſe bars on the head, neck, and back are red and black; the breaſt is of a pale orange colour; belly barred with orange and black, the top of each feather being white; the back and wings are mottled with reddiſh brown and black, the ſcapulars tipped with white; the tail is of a deep ruſt colour, barred with black, and tipped with white.

This beautiful kind is found chiefly in high mountainous regions, and is very rare in Great Britain. Mr Pennant mentions one, which was ſhot near Inverneſs, as an uncommon inſtance. It was formerly met with in Ireland, but is now ſuppoſed to be extinct there. In Ruſſia, Sweden, and other northern countries, it is very common: It lives in the foreſts of pine, with which thoſe countries abound, and feeds on the cones of the fir trees, which, at ſome ſeaſons, give an unpleaſant flavour to its fleſh ſo as to render it unfit for the table; it likewiſe eats various kinds of plants and berries, particularly the juniper. Early in the ſpring the ſeaſon of pairing commences: During this period, the cock places himſelf on an eminence, where he diſplays a variety of pleaſing attitudes; the feathers on his head ſtand erect, his neck ſwells, his tail is diſplayed, and his wings trail almoſt on the ground, his eyes ſparkle, and the ſcarlet patch on each ſide of his head aſſumes a deeper dye; at the ſame time he utters his well-known cry, which has been compared to the ſound produced by the whetting of a ſcythe; it may be heard at a conſiderable diſtance, and never fails to draw around him his faithful mates. The female lays from eight to ſixteen eggs, which are white, ſpotted with yellow, and larger than thoſe of the common Hen; for this purpoſe ſhe chuſes ſome ſecret ſpot, where ſhe can ſit in ſecurity; ſhe covers her eggs carefully over with leaves, when ſhe is under the neceſſity of leaving them in ſearch of food. The young follow the hen as ſoon as they are hatched, ſometimes with part of the ſhell attached to them.

THE BLACK GROUSE.

BLACK GAME, OR BLACK COCK.

(*Tetrao Tetrix*, Lin.—*Le Coq de Bruyere a queue four-chue*, Buff.)

This bird, though not larger than a fowl, weighs near four pounds; its length is about one foot ten inches; breadth two feet nine: The bill is black; the eyes dark blue; below each eye there is a ſpot of a dirty white colour, and above a larger one, of a bright ſcarlet, which extends almoſt to the top of the head; the general colour of the plumage is of a deep black, richly gloſſed with blue on the neck and rump; the leſſer wing coverts are duſky brown; the greater are white, which extends to the ridge of the wing, forming a ſpot of that colour on the ſhoulder when the wing is cloſed; the quills are brown, the lower parts and tips of the ſecondaries are white, forming a bar of white acroſs the wing—there is likewiſe a ſpot of white on the baſtard wing; the feathers of the tail are almoſt ſquare at the ends, and when ſpread out, form a curve on each ſide; the under tail coverts are of a pure white; the legs and thighs are of a dark brown colour, mottled with white; the toes are toothed on the edges like the former ſpecies. In ſome of our ſpecimens the noſtrils were thickly covered with feathers, whilſt in others they were quite bare, which we ſuppoſe muſt be owing to the different ages of the birds.

Theſe birds, like the former, are found chiefly in high and wooded ſituations in the northern parts of our iſland; they are common in Ruſſia, Siberia, and other northern countries: They feed on various kinds of berries and other fruits, the produce of wild and mountainous places; in ſummer they will frequently come down from their lofty ſituations for the ſake of feeding on corn. They do not pair, but on the return of ſpring the males aſſemble in great numbers at their accuſtomed reſorts, on the tops of high and heathy mountains, when the conteſt for ſuperiority takes place, and continues with great bitterneſs till the vanquiſhed are put to flight; the victors being left in poſſeſſion of the field, place themſelves on an eminence, clap their wings, and with loud cries give notice to their females, who immediately reſort to the ſpot. It is ſaid that each cock has two or three hens, which ſeem particularly attached to him. The female is about one-third leſs than the male, and differs conſiderably in colour; her tail is likewiſe much leſs forked: She makes an artleſs neſt on the ground, where ſhe lays ſix or eight eggs, of a yellowiſh colour, with freckles and ſpots of a ruſty brown: The young males at firſt reſemble the mother; they do not acquire their full plumage till toward the end of autumn, when it gradually changes to a deeper colour, and aſſumes that of a blueiſh black, which it afterwards retains.

RED GROUSE.

RED GAME, GORCOCK, OR MOORCOCK.

(*Tetrao Scoticus*, Lin.—*L'Attagas*, Buff.)

The length of this bird is fifteen inches; the weight about nineteen ounces: The bill is black; the eyes hazel; the noſtrils ſhaded with ſmall red and black feathers; at the baſe of the lower bill there is a white ſpot on each ſide; the throat is red; each eye is arched with a large naked ſpot, of a bright ſcarlet colour; the whole upper part of the body is beautifully mottled with deep red and black, which gives it the appearance of tortoiſe-ſhell; the breaſt and belly are of a purpleiſh hue, croſſed with ſmall duſky lines; the tail conſiſts of ſixteen feathers, of equal lengths, the four middle-moſt barred with red, the others black; the quills are duſky; the legs are clothed with ſoft white feathers down to the claws, which are ſtrong, and of a light colour. The female is ſomewhat leſs; the naked ſkin above each eye is not ſo conſpicuous, and the colours of its plumage in general much lighter than thoſe of the male.

This bird is found in great plenty in the wild, heathy, and mountainous tracts in the northern counties of England; it is likewiſe common in Wales, and in the highlands of Scotland. Mr Pennant ſuppoſes it to be peculiar to Britain; thoſe found in the mountainous parts of France, Spain, Italy, and elſewhere, as mentioned by M. Buffon, are probably varieties of this kind, and we have no doubt would breed with it. We. could wiſh that attempts were more frequently made to introduce a greater variety of theſe uſeful birds into this country, to ſtock our waſte and barren moors with a rich fund of delicate and wholeſome food; but, till a wiſe and enlightened legiſlature ſhall alter or abrogate our very unequal and injudicious game laws, there hardly remains a ſingle hope for the preſervation of thoſe we have. Red Grouſe pair in the ſpring; the female lays eight or ten eggs, on the ground: The young ones follow the hen the whole ſummer; as ſoon as they have attained their full ſize, they unite in flocks of forty or fifty, and are then exceedingly ſhy and wild.

WHITE GROUSE.

WHITE GAME, OR PTARMIGAN.

(*Tetrao lagopus*, Lin.—*Le Lagopède*, Buff.)

THIS bird is nearly the ſame ſize as the Red Grouſe: Its bill is black; the upper parts of its body are of a pale brown or aſh colour, mottled with ſmall duſky ſpots and bars; the bars on the head and neck are ſomewhat broader, and are mixed with white; the under parts are white, as are alſo the wings, excepting the ſhafts of the quills, which are black. This is its ſummer dreſs; in winter it changes to a pure white, except that in the male there is a black line between the bill and the eye; the tail conſiſts of ſixteen feathers; the two middle ones are aſh-coloured in ſummer, and white in winter, the two next ſlightly marked with white near the ends, the reſt are wholly black; the upper tail coverts are long, and almoſt cover the tail.

The White Grouſe is fond of lofty ſituations, where it braves the ſevereſt cold: It is found in moſt of the northern parts of Europe, even as far as Greenland; in this country it is only to be met with on the ſummits of ſome of our higheſt hills, chiefly in the highlands of Scotland, in the Hebrides and Orkneys, and ſometimes, but rarely, on the lofty hills of Cumberland and Wales. Buffon, ſpeaking of this bird, ſays, that it avoids the ſolar heat, and prefers the biting froſts on the tops of mountains; for, as the ſnow melts on the ſides of the mountains, it conſtantly aſcends, till it gains the ſummit, where it forms holes, and burrows in the ſnow. They pair at the ſame time with the Grouſe; the female lays eight or ten eggs, which are white, ſpotted with brown; ſhe makes no neſt, but depoſits them on the ground. In winter they fly in flocks, and are ſo little accuſtomed to the ſight of man, that they ſuffer themſelves to be eaſily taken either with the ſnare or gun. They feed on the wild productions of the hills, which ſometimes give the fleſh a bitter taſte, but not unpalateable; it is dark coloured, and according to M. Buffon has ſomewhat the flavour of hare.

THE PARTRIDGE.

(*Tetrao perdix*, Lin.—*Le perdrix Grife*, Buff.)

The length of this bird is about thirteen inches: The bill is light brown; eyes hazel; the general colour of its plumage is brown and afh, elegantly mixed with black, each feather being ftreaked down the middle with buff colour; the fides of the head are tawny; under each eye there is a fmall faffron-coloured fpot, which has a granulated appearance, and between the eye and the ear a naked fkin of a bright fcarlet, which is not very confpicuous but in old birds; on the breaft there is a crefcent of a deep cheftnut colour; the tail is fhort; the legs are of a greenifh white, and are furnifhed with a fmall knob behind: The female wants the creſcent on the breaſt, and its colours in general are not ſo diſtinct and bright.

Partridges are chiefly found in temperate climates, the extremes of heat and cold being equally unfavourable to them: They are no where in greater plenty than in this iſland, where, in their ſeaſon, they contribute to our moſt elegant entertainments. It is much to be lamented, however, that the means taken to preſerve this valuable bird ſhould, in a variety of inſtances, prove its deſtruction; the proper guardians of the young ones and eggs, tied down by ungenerous reſtrictions, are led to conſider them as a growing evil, and not only connive at their deſtruction, but too frequently aſſiſt in it. Partridges pair early in the ſpring; the female lays from fourteen to eighteen or twenty eggs, making her neſt of dry leaves and graſs upon the ground: The young birds learn to run as ſoon as hatched, frequently encumbered with part of the ſhell ſticking to them. It is no uncommon thing to introduce Partridge eggs under the common Hen, who hatches and rears them as her own: In this caſe the young birds require to be fed with ants' eggs, which is their favourite food, and without which it is almoſt impoſſible to bring them up; they likewiſe eat inſects, and, when full grown, feed on all kinds of grain and young plants. The affection of the Partridge for her young is peculiarly ſtrong and lively; ſhe is greatly aſſiſted in the care of rearing them by her mate; they lead them out in common, call them together, point out to them their proper ſood, and aſſiſt them in finding it by ſcratching the ground with their feet; they frequently ſit cloſe by each other, covering the chickens with their wings, like the Hen: In this ſituation they are not eaſily fluſhed; the ſportſman, who is attentive to the preſervation of his game, will carefully avoid giving any diſturbance to a ſcene ſo truly intereſting; but ſhould the pointer come too near, or unfortunately run in upon them, there are few who are ignorant of the confuſion that follows: The male firſt gives the ſignal of alarm by a peculiar cry of diſtreſs, throwing himſelf at the ſame moment more immediately into the way of danger, in order to deceive or miſlead the enemy; he flies, or rather runs along the ground, hanging his wings, and exhibiting every ſymptom of debility, whereby the dog is decoyed, by a too eager expectation of an eaſy prey, to a diſtance from the covey; the female flies off in a contrary direction, and to a greater diſtance, but returning ſoon after by ſecret ways, ſhe finds her ſcattered brood cloſely ſquatted among the graſs, and collecting them with haſte, ſhe leads them from the danger, before the dog has had time to return from his purſuit.

THE QUAIL.

(*Tetrao coturnix*, Lin.—*La Caille*, Buff.)

The length ſeven inches and a half: Bill duſky; eyes hazel; the colours of the head, neck, and back are a mixture of brown, aſh colour, and black; over each eye there is a yellowiſh ſtreak, and another of the ſame colour down the middle of the forehead; a dark line paſſes from each corner of the bill, forming a kind of gorget above the breaſt; the ſcapular feathers are marked by a light yellowiſh ſtreak down the middle of each; the quills are of a lightiſh brown, with ſmall ruſt coloured bands on the exterior edges of the feathers; the breaſt is of a pale ruſt colour, ſpotted with black, and ſtreaked with pale yellow; the tail conſiſts of twelve feathers, barred like the wings; the belly and thighs are of a yellowiſh white; legs pale brown. We were favoured with a very fine ſpecimen of this beautiful bird alive by Mr Gilfrid Ward, and one ſhot by the Rev. Mr Brocklebank of Corbridge, from which our repreſentation was made. The female wants the black ſpots on the breaſt, and is eaſily diſtinguiſhed by a leſs vivid plumage.

Quails are almoſt univerſally diffuſed throughout Europe, Aſia, and Africa; they are birds of paſſage, and are ſeen in immenſe flocks traverſing the Mediterranean ſea from Italy to the ſhores of Africa in the autumn, and returning again in the ſpring, frequently alighting in their paſſage on many of the iſlands of the Archipelago, which they almoſt cover with their numbers. On the weſtern coaſts of the kingdom of Naples ſuch prodigious quantities have appeared, that an hundred thouſand have been taken in a day within the ſpace of four or five miles. From theſe circumſtances it appears highly probable, that the Quails which ſupplied the Iſraelites with food, during their journey through the wilderneſs, were ſent thither on their paſſage to the north by a wind from the ſouth-weſt, ſweeping over Egypt and Ethiopia towards the ſhores of the Red ſea. Quails are not very numerous in this iſland; they breed with us, and many of them remain throughout the year, changing their quarters from the interior counties to the ſea coaſt. The female makes her neſt like the Partridge, and lays to the number of ſix or ſeven* eggs, of a greyiſh colour, ſpeckled with brown: The young birds follow the mother as ſoon as hatched, but, do not continue long together; they are ſcarcely grown up before they ſeparate; or, if kept together, they fight obſtinately with each other, their quarrels frequently terminating in each other's deſtruction. From this quarrelſome diſpoſition in the Quail it was, that they were formerly made uſe of by the Greeks and Romans, as we do Game-cocks, for the purpoſe of fighting. We are told that Auguſtus puniſhed a prefect of Egypt with death, for bringing to his table one of theſe birds which had acquired celebrity by its victories. At this time the Chineſe are much addicted to the amuſement of fighting Quails, and in ſome parts of Italy it is ſaid likewiſe to be no unuſual practice. After feeding two Quails very highly, they place them oppoſite to each other, and throw in a few grains of ſeed between them—the birds ruſh upon each other with the utmoſt fury, ſtriking with their bills and ſpurs till one of them yields.

* In France they are ſaid to lay fifteen or twenty.—*Buff.*

THE CORN-CRAKE.

LAND RAIL, OR DAKER HEN.

(*Rallus-Crex*, Lin.—*Le Rale de genet*, Buff.)

Length rather more than nine inches: The bill is light brown; the eyes hazel; all the feathers on the upper parts of the plumage are of a dark brown, edged with pale ruſt colour; both wing coverts and quills are of a deep cheſtnut; the fore part of the neck and breaſt is of a pale aſh colour; a ſtreak of the ſame colour extends over each eye from the bill to the ſide of the neck; the belly is of a yellowiſh white; the ſides, thighs, and vent are faintly marked with ruſty coloured ſtreaks; the legs are of a pale fleſh colour.

We have ventured to remove this bird from the uſual place aſſigned to it among thoſe to which it ſeems to have little or no analogy, and have placed it among others, to which, in moſt reſpects, it bears a ſtrong affinity. It makes its appearance about the ſame time with the Quail, and frequents the ſame places, from whence it is called, in ſome countries, the king of the Quails. Its well-known cry is firſt heard as ſoon as the graſs becomes long enough to afford it ſhelter, and continues till the time it is cut, but is ſeldom ſeen; it conſtantly ſkulks among the thickeſt part of the herbage, and runs ſo nimbly through it, winding and doubling in every direction, that it is difficult to come near it; when hard puſhed by the dog, it ſometimes ſtops ſhort and ſquats down, by which means, its too eager purſuer overſhoots the ſpot, and loſes the trace. It ſeldom ſprings but when driven to extremity, and generally flies with its legs hanging down, but never to a great diſtance: As ſoon as it alights it runs off, and before the fowler has reached the ſpot, the bird is at a conſiderable diſtance. —The Corn-crake leaves this iſland in winter, and repairs to other countries in ſearch of food, which conſiſts of worms, ſlugs, and inſects; it likewiſe feeds on ſeeds of various kinds: It is very common in Ireland, and is ſeen in great numbers in the iſland of Angleſea in its paſſage to that country. On its firſt arrival in England it is ſo lean as to weigh leſs than ſix ounces, from whence one would conclude that it muſt have come from diſtant parts; before its departure, however, it has been known to exceed eight ounces, and is then very delicious eating. The female lays ten or twelve eggs, on a neſt made of a little moſs or dry graſs careleſsly put together; they are of a pale aſh colour, marked with ruſt-coloured ſpots. The young Crakes run as ſoon as they have burſt the ſhell, following the mother; they are covered with a black down, and ſoon find the uſe of their legs.—Our figure was made from the living bird ſent us by Lieut. H. F. Gibſon.

GREAT BUSTARD.

(*Otis tarda*, Lin.—*L'Outarde*, Buff.)

This very ſingular bird, which is the largeſt of our land birds, is about four feet long, and weighs from twenty-five to thirty pounds; its characters are peculiar, and with thoſe which connect it with birds of the gallinaceous kind, it has others which ſeem to belong to the Oſtrich and the Caſſowary: Its bill is ſtrong and rather convex; its eyes red; on each ſide of the lower bill there is a tuft of feathers about nine inches long; its head and neck are aſh-coloured. In the one deſcribed by Edwards, there were on each ſide of the neck two naked ſpots, of a violet colour, but which appeared to be covered with feathers when the neck was much extended. The back is barred tranſverſely with black and bright ruſt colour; the quills are black; the belly white; the tail conſiſts of twenty feathers —the middle ones are ruſt colour, barred with black; thoſe on each ſide are white, with a bar or two of black near the ends; the legs are long, naked above the knees, and duſky; it has no hind toe; its nails are ſhort, ſtrong, and convex both above and below; the bottom of the foot is furniſhed with a callous prominence, which ſerves inſtead of a heel.—The female is not much more than half the ſize of the male: The top of her head is of a deep orange, the reſt of the head brown; her colours are not ſo bright as thoſe of the male, and ſhe wants the tuft on each ſide of the head: There is likewiſe another very eſſential difference between the male and the female, the former being furniſhed with a ſac or pouch, which is ſituated in the fore part of the neck, and is capable of containing about two quarts; the entrance to it is immediately under the tongue.* This ſingular reſervoir was firſt diſcovered by Dr Douglas, who ſuppoſes that the bird fills it with water as a ſupply in the midſt of thoſe dreary plains where it is accuſtomed to wander; it likewiſe makes a further uſe of it in defending itſelf againſt the attack of birds of prey; on ſuch occaſions it throws out the water with ſuch violence, as not unfrequently to baffle the purſuit of its enemy.

Buſtards were formerly more frequent in this iſland than at preſent; they are now found only in the open countries of the South and Eaſt, in the plains of Wiltſhire, Dorſetſhire, and in ſome parts of Yorkſhire; they were formerly met with in Scotland, but are now ſuppoſed to be extinct there. They are ſlow in taking wing, but run with great rapidity, and when young are ſometimes taken with greyhounds, which purſue them with great avidity: The chace is ſaid to afford excellent diverſion. The Great Buſtard is granivorous, feeding on herbs and grain of various kinds; it is alſo fond of thoſe worms which are ſeen to come out of the ground in great numbers before ſun-riſe in the ſummer; in winter it frequently feeds on the bark of trees: Like the Oſtrich, it ſwallows ſmall ſtones,†

* Barrington's Miſ. p. 553.

† In the ſtomach of one which was opened by the academicians there were found, beſides ſmall ſtones, to the number of ninety doubloons, all worn and poliſhed by the attrition of the ſtomach.—*Buff.*

bits of metal, and the like. The female makes no neſt, but, making a hole on the ground, ſhe drops two eggs, about the ſize of thoſe of a Gooſe, of a pale olive brown, with dark ſpots: She ſometimes leaves her eggs in queſt of food; and if during her abſence, any one ſhould handle, or even breathe upon them, ſhe immediately abandons her neſt. Buſtards are found in various parts of Europe, Aſia, and Africa, but have not hitherto been diſcovered on the new continent.

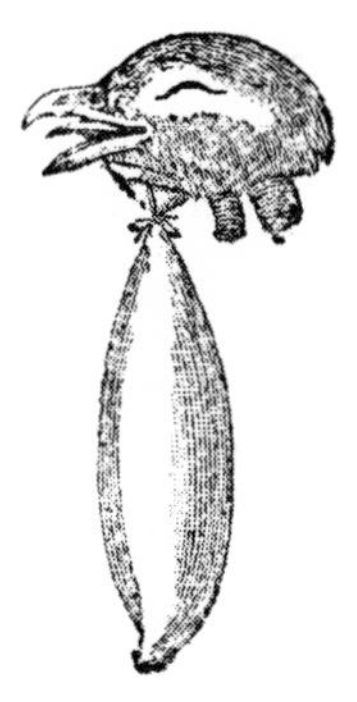

LITTLE BUSTARD.

(*Otis Tetrax*, Lin.—*Le petite Outarde*, Buff.)

Length only ſeventeen inches: The bill is pale brown; eyes red; the top of the head is black, ſpotted with pale ruſt colour; the ſides of the head, chin, and throat, are of a reddiſh white, marked with a few dark ſpots; the whole neck is black, encircled with an irregular band of white near the top and bottom; the back and wings are ruſt colour, mottled with brown, and croſſed with fine irregular black lines; the under parts of the body, and outer edges of the wings are white; the tail conſiſts of eighteen feathers—the middle ones are tawny, barred with black, the others are white, marked with a few irregular bands of black; the legs are gray. The female is ſmaller, and wants the black collar on its neck; in other reſpects ſhe nearly reſembles the male.

This bird is very uncommon in this country; we have ſeen only two of them, both of which were females: Our figure was taken from one ſent us by W. Trevelyan, Eſq. which was taken on the edge of Newmarket Heath, and kept alive about three weeks, in a kitchen, where it was fed with bread, and other things, ſuch as poultry eat. Both this and the Great Buſtard are excellent eating, and, we would imagine, would well repay the trouble of domeſtication; indeed it ſeems ſurpriſing that we ſhould ſuffer theſe fine birds to run wild, and be in danger of total extinction, which, if properly cultivated, might afford as excellent a repaſt as our own domeſtic poultry, or even the Turkey, for which we are indebted to diſtant countries: It is very common in France, where it is taken in nets like the Partridge: It is a very ſhy and cunning bird; if diſturbed, it flies two or three hundred paces, not far from the ground, and then runs away much faſter than any one can follow on foot. The female lays her eggs in June to the number of three or four, of a gloſſy green colour; as ſoon as the young are hatched, ſhe leads them about as the hen does her chickens; they begin to fly about the middle of Auguſt.

OF THE PLOVER.

Although the Plover has generally been claſſed with thoſe birds whoſe buſineſs is wholly among waters, we cannot help conſidering the greater part of them as partaking entirely of the nature of land birds. Many of them breed upon our loftieſt mountains, and though they are frequently ſeen upon the ſea-coaſts, feeding with birds of the water kind, yet it muſt be obſerved that they are no more water birds than many of our ſmall birds who repair there for the ſame purpoſe.

The Plover is diſtinguiſhed by a large full eye; its bill is ſtrait, ſhort, and rather ſwelled towards the tip; its head is large; and its legs are naked above the knee. The long-legged Plover and the Sanderling are waders, and belong more immediately to the water birds, to which we refer them: The Great Plover and the Lapwing we conſider as entirely connected with birds of the Plover kind; the former has uſually been claſſed with the Buſtard, the latter with the Sandpiper; but they differ very materially from both, and ſeem to agree in more eſſential points with this kind: We have therefore given them a place in this part of our work, where they may be conſidered as connecting the two great diviſions of land and water birds, to both of which they are in ſome degree allied.

THE GREAT PLOVER.

THICK-KNEE'D BUSTARD, STONE CURLEW, NORFOLK PLOVER.

(*Charadrius Oedicnemus*, Lin.—*Le grand Pluvier*, Buff.)

The length of this bird is about ſixteen inches; Its bill is long, yellowiſh at the baſe, and black at the end; its eyes and eye-lids are pale yellow; above each eye there is a pale ſtreak, and beneath one of the ſame colour extends to the bill; the throat is white; head, neck, and all the upper parts of the body are of a pale tawny brown, down the middle of each feather there is a dark ſtreak; the fore part of the neck and breaſt are nearly the ſame, but much paler; the belly, thighs, and vent are of a pale yellowiſh white; the quills are black; the tail is ſhort and rounded—a dark band croſſes the middle of each feather, the tips are black, the reſt white; the legs are yellow, and naked above the knees, which are very thick, as if ſwelled—hence its name; the claws are black.

This bird is found in great plenty in Norfolk and ſeveral of the ſouthern counties, but is no where to be met with in the northern parts of our iſland; it prefers dry and ſtony places, on the ſides of ſloping banks: It makes no neſt; the female lays two or three eggs on the bare ground, ſheltered by a ſtone, or in a ſmall hole formed in the ſand; they are of a dirty white, marked with ſpots of a deep reddiſh colour, mixed with ſlight ſtreaks. Although this bird has great power of wing, and flies with great ſtrength, it is ſeldom ſeen during the day, except ſurpriſed, when it ſprings to ſome diſtance, and generally eſcapes before the ſportſman comes within gun-ſhot; it likewiſe runs on the ground almoſt as ſwift as a dog; after running ſome time it ſtops ſhort, holding its head and body ſtill, and on the leaſt noiſe ſquats cloſe on the ground. In the evening it comes out in queſt of food, and may then be heard at a great diſtance; its cry is ſingular, reſembling a hoarſe kind of whiſtle three or four times repeated, and has been compared to the turning of a rufty handle. Buffon endeavours to exprefs it by the words ***turrlui, turrlui,*** and fays, it refembles the found of a third flute, dwelling on three or four tones from a flat to a fharp. Its food confifts chiefly of worms. It is faid to be good eating when young; the flefh of the old ones is hard, black, and dry. Mr White mentions them as frequenting the diftrict of Selborne, in Hampfhire. He fays, that the young run immediately from the neft, almoft as foon as they are excluded, like Partridges; that the dam leads them to fome ftony field, where they bafk, fkulking among the ftones, which they refemble fo nearly in colour, as not eafily to be difcovered. Birds of this kind are migratory; they arrive in April, live with us all the fpring and fummer, and at the beginning of autumn prepare to take leave by getting together in flocks; it is fuppofed that they retire to Spain, and frequent the fheep-walks with which that country abounds.

THE PEE-WIT.

LAPWING, BASTARD PLOVER, OR TE-WIT.

(*Fringilla vanellus*, Lin.—*Le Vanneau*, Buff.)

THIS bird is about the ſize of a Pigeon: Its bill is black; eyes large and hazel; the top of the head is black, gloſſed with green; a tuft of long narrow feathers iſſues from the back part of the head, ſome of which are four inches in length, and turn upwards at the end; the ſides of the head and neck are white, which is interrupted by a blackiſh ſtreak above and below the eye; the back part of the neck is of a very pale brown; the fore part, as far as the breaſt, is black; the back and wing coverts are of a dark green, gloſſed with purple and blue reflections; the quills are black, the four firſt tipped with white; the breaſt and belly are of a pure white; the upper tail coverts and vent pale cheſtnut; the tail is white at the baſe, the end is black, with pale tips, the outer feathers almoſt wholly white; the legs are red; claws black; hind claw very ſhort.

This bird is a conſtant inhabitant of this country; but as it ſubſiſts chiefly on worms, it is forced to change its place in queſt of food, and is frequently ſeen in great numbers by the ſea-ſhores, where it finds an abundant ſupply. It is every where well known by its loud and inceſſant cries, which it repeats without intermiſſion, whilſt on the wing, and from whence, in moſt languages, a name has been given to it as imitative of the ſound.—The Pee-wit is a lively active bird, almoſt continually in motion; it ſports and frolics in the air in all directions, and aſſumes a variety of attitudes; it remains long upon the wing, and ſometimes riſes to a conſiderable height; it runs along the ground very nimbly, and ſprings and bounds from ſpot to ſpot with great agility: The female lays four eggs, of a dirty olive, ſpotted with black; ſhe makes no neſt, but depoſits them upon a little dry graſs haſtily ſcraped together; the young birds run very ſoon after they are hatched;—during this period the old ones are very aſſiduous in their attention to their charge; on the approach of any perſon to the place of their depoſit, they flutter round his head with cries of the greateſt inquietude, which increaſes as he draws nearer the ſpot where the brood are ſquatted; in caſe of extremity, and as a laſt reſource, they run along the ground as if lame, in order to draw off the attention of the fowler from any further purſuit. The young Lapwings are firſt covered with a blackiſh down interſperſed with long white hairs, which they gradually loſe, and about the latter end of July they acquire their beautiful plumage. At this time, according to Buffon, the great aſſociation begins to take place, and they aſſemble in large flocks of young and old, which hover in the air, ſaunter in the meadows, and after rain they diſperſe among the ploughed fields. In the month of October the Lapwings are very fat, and are then ſaid to be excellent eating: Their eggs are conſidered as a great delicacy, and are ſold in the London markets at three ſhillings a dozen.

The following anecdote, communicated to us by the Rev. J. Carlyle, is worthy of notice, as it ſhews the domeſtic nature of this bird, as well as the art with which it conciliates the regard of animals differing from itſelf in nature, and generally conſidered as hoſtile to every ſpecies of the feathered tribes. Two of theſe birds, given to Mr Carlyle, were put into a garden, where one of them ſoon died; the other continued to pick up ſuch food as the place afforded, till winter deprived it of its uſual ſupply; neceſſity ſoon compelled it to draw nearer the houſe, by which it gradually became familiariſed to occaſional interruptions from the family. At length, one of the ſervants, when ſhe had occaſion to go into the back-kitchen with a light, obſerved that the Lapwing always uttered his cry '*pee-wit*' to obtain admittance. He ſoon grew more familiar; as the winter advanced, he approached as far as the kitchen, but with much caution, as that part of the houſe was generally occupied by a dog and a cat, whoſe friendſhip the Lapwing at length conciliated ſo entirely, that it was his regular cuſtom to reſort to the fireſide as ſoon as it grew dark, and ſpend the evening and night with his two aſſociates, ſitting cloſe by them, and partaking of the comforts of a warm fireſide. As ſoon as ſpring appeared, he left off coming to the houſe, and betook himſelf to the garden; but on the approach of winter, he had recourſe to his old ſhelter and his old friends, who received him very cordially. Security was productive of inſolence; what was at firſt obtained with caution, was afterwards taken without reſerve: He frequently amuſed himſelf with waſhing in the bowl which was ſet for the dog to drink out of, and while he was thus employed, he ſhewed marks of the greateſt indignation if either of his companions preſumed to interrupt him. He died in the aſylum he had chofen, being choaked with fomething which he picked up from the floor. During his confinement, crumbs of wheaten bread were his principal food, which he preferred to any thing elfe.

THE GOLDEN PLOVER.

YELLOW PLOVER.

(*Charadrius Pluvialis*, Lin.—*Le Pluvier doré*, Buff.)

The ſize of the Turtle: Bill duſky; eyes dark; all the upper parts of the plumage are marked with bright yellow ſpots upon a dark brown ground; the fore part of the neck and breaſt are the ſame, but much paler; the belly is almoſt white; the quills are duſky; the tail is marked with duſky and yellow bars; the legs are black.—Birds of this ſpecies vary much from each other; in ſome which we have had, the breaſt was marked with black and white; in others, it was almoſt black; but whether this difference aroſe from age or ſex, we are at a loſs to determine.

The Golden Plover is common in this country, and all the northern parts of Europe; it is very numerous in various parts of America, from Hudſon's Bay as far as Carolina, migrating from one place to another according to the ſeaſons: It breeds on high and heathy mountains; the female lays four eggs, of a pale olive colour, variegated with blackiſh ſpots: They fly in ſmall flocks, and make a ſhrill whiſtling noiſe, by an imitation of which they are ſometimes enticed within gun-ſhot. The male and female do not differ from each other. In young birds the yellow ſpots are not very diſtinguiſhable, the plumage inclining more to gray.

THE GRAY PLOVER.

(*Tringa Squatarola*, Lin.—*Le vanneau Pluvier*, Buff.)

The length of this bird is about twelve inches: Its bill is black; the head, back, and wing coverts are of a dufty brown, edged with greenifh afh colour, and fome with white; the cheeks and throat are white, marked with oblong dufky fpots; the belly, thighs, and rump are white; the fides are marked with a few dufky fpots; the outer webs of the quills are black, the lower parts of the inner webs of the four firft are white; the tail is marked with alternate bars of black and white; the legs are of a dull green; its hind toe is fmall.—In the *Planches Enluminees* this bird is reprefented with eyes of an orange colour; there is likewife a dufky line extending from the bill underneath each eye, and a white one above it.

We have placed this bird with the Plovers, as agreeing with them in every other refpect but that of having a hind toe; but that is fo fmall as not to render it neceffary to exclude it from a place in the Plover family, to which it evidently belongs. The Gray Plover is not very common in Britain; it appears fometimes in fmall flocks on the fea-coafts: It is fomewhat larger than the Golden Plover. Its flefh is faid to be very delicate.

THE DOTTEREL.

(*Charadrius Morinellus*, Lin.—*Le Guignard*, Buff.)

The length of this bird is about nine inches: Its bill is black; eyes dark, large, and full; its forehead is mottled with brown and white; top of the head black; over each eye an arched line of white paſſes to the hind part of the neck; the cheeks and throat are white; the back and wings are of a light brown, inclining to olive, each feather being margined with pale ruſt colour; the quills are brown; the fore part of the neck is ſurrounded by a broad band of a light olive colour, bordered on the under ſide with white; the breaſt is of a pale dull orange; middle of the belly black; the reſt of the belly, thighs, and vent are of a reddiſh white; the tail is of an olive brown, black near the end, and tipped with white—the outer feathers are margined with white; the legs are of a dark olive colour.

The Dotterel is common in various parts of Great Britain; in others it is ſcarcely known:—They are ſuppoſed to breed in the mountains of Cumberland and Weſtmorland, where they are ſometimes ſeen in the month of May, during the breeding ſeaſon; they likewiſe breed on ſeveral of the Highland hills: They are very common in Cambridgeſhire, Lincolnſhire, and Derbyſhire, appearing in ſmall flocks on the heaths and moors of thoſe counties during the months of May and June, and are then very fat, and much eſteemed for the table. The Dotterel is ſaid to be a very ſtupid bird, and eaſily taken with the moſt ſimple artifice, and that it was formerly the cuſtom to decoy them into the net by ſtretching out a leg or an arm, which caught the attention of the birds, ſo that they returned it by a ſimilar motion of a leg or a wing, and were not aware till the net dropped and covered the whole covey. At preſent the more ſure method of the gun has ſuperſeded this ingenious artifice.

THE RING DOTTEREL.

RING PLOVER, OR SEA LARK.

(*Charadrius Hiaticula*, Lin.—*Le petit Pluvier à collier*, Buff.)

The length is rather more than ſeven inches: The bill is of an orange colour, tipped with black; the eyes are hazel; a black line paſſes from the bill, underneath each eye, to the cheeks, where it is pretty broad; above this a line of white extends acroſs the forehead to the eyes—this is bounded above by a black fillet acroſs the head; a gorget of black encircles the neck, very broad on the fore part, but growing narrow behind—above which, to the chin, is white; the top of the head is of a light brown aſh colour, as are alſo the back, ſcapulars, and coverts; the greater coverts are tipped with white; the breaſt and all the under parts are white ; the quills are dufky, with an oval white fpot about the middle of each feather, which forms, when the wing is clofed, a ftroke of white down each wing ; the tail is of a dark brown, tipped with white, the two outer feathers almoft white ; the legs are of an orange colour ; claws black.—In the female, the white on the forehead is lefs ; there is more white on the wings, and the plumage inclines more to afh colour.

Thefe birds are common in all the northern countries ; they migrate into Britain in the fpring, and depart in autumn : They frequent the fea-fhores during fummer ; they run nimbly along the fands, fometimes taking fhort flights, accompanied with loud twitterings, then alight and run again : If difturbed, they fly quite off. They are faid to make no neft ; the female lays four eggs, of a pale afh colour, fpotted with black, which fhe depofits on the ground.

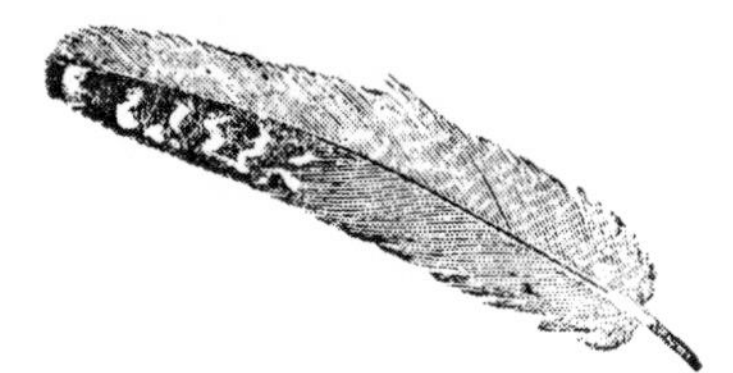